Fundamentals of Calculus

Fundamentals of Calculus

Gerald Atkins

Larsen & Keller
www.larsen-keller.com

Fundamentals of Calculus
Gerald Atkins
ISBN: 978-1-64172-140-0 (Hardback)

Larsen & Keller
Published by Larsen and Keller Education,
5 Penn Plaza,
19th Floor,
New York, NY 10001, USA

Cataloging-in-Publication Data

Fundamentals of calculus / Gerald Atkins.
p. cm.
Includes bibliographical references and index.
ISBN 978-1-64172-140-0
1. Calculus. 2. Mathematical analysis. 3. Functions.
4. Mathematical analysis. I. Atkins, Gerald.
QA303.2 .F85 2019
515--dc23

For more information regarding Larsen and Keller Education and its products, please visit the publisher's website www.larsen-keller.com

Table of Contents

Preface

Calculus is a branch of mathematics that studies continuous change. It can be divided into the two branches of differential and integral calculus. The principles of limits and infinitesimals, the fundamental theorem of calculus and the convergence of infinite sequences and infinite series are fundamental to the development of calculus. Current studies in this field are in the areas of reformulations of calculus such as non-standard calculus, smooth infinitesimal analysis and constructive analysis. An understanding of this domain is crucial for developing a functional approach to advanced mathematical analysis. Besides advancing the frontiers of advanced mathematics, calculus is also instrumental in science, engineering and economics. This book provides comprehensive insights into the field of calculus. Some of the diverse topics covered herein address the varied branches that fall under this category. Coherent flow of topics, student-friendly language and extensive use of examples make this textbook an invaluable source of knowledge.

A detailed account of the significant topics covered in this book is provided below:

Chapter 1- Calculus is the mathematical study of continuous change. It is divided into two major branches, namely differential calculus and integral calculus. The various concepts elucidated in this chapter include the core topics of limits, continuous functions, fundamental theorem of calculus and geometric calculus. It closely examines these key concepts and branches of calculus to provide an extensive understanding of the subject.

Chapter 2- Differential calculus is an important sub-field of calculus, which studies the rates of change of quantities. The topics of derivatives and differential are integral to the study of differential calculus. This chapter discusses the fundamental concepts of differential calculus such as differentiation rules, differentiable functions and implicit functions.

Chapter 3- Integral calculus is another significant sub-domain of calculus. Integration is the inverse operation of differentiation. This chapter discusses in detail the different kinds of integrals such as, surface integral, multiple integral, Frullani integral, etc. It has been carefully written to provide an easy understanding of the varied aspects of integral calculus.

Chapter 4- Vector calculus is a branch of mathematics. It is concerned with the differentiation and integration of vector fields in 3D Euclidean space. The significant concepts of vector operators like the del, gradient, divergence and curl operator have been discussed in detail in this chapter. It also closely examines the key theories of vector calculus such as gradient theorem, divergence theorem, Stokes' theorem and Green's theorem, to provide an extensive understanding of the subject.

It gives me an immense pleasure to thank our entire team for their efforts. Finally in the end, I would like to thank my family and colleagues who have been a great source of inspiration and support.

Gerald Atkins

1

Calculus: An Introduction

Calculus is the mathematical study of continuous change. It is divided into two major branches, namely differential calculus and integral calculus. The various concepts elucidated in this chapter include the core topics of limits, continuous functions, fundamental theorem of calculus and geometric calculus. It closely examines these key concepts and branches of calculus to provide an extensive understanding of the subject.

Calculus is a branch of mathematics that explores variables and how they change by looking at them in infinitely small pieces called Infinitesimals. Calculus, as it is practiced today, was invented by British scientist Isaac Newton (1642 to 1726) and German scientist Gottfried Leibnitz (1646 to 1716), who independently developed the principles of calculus in the traditions of geometry and symbolic mathematics, respectively.

While these two discoveries are most important to calculus as it is practiced today, they were not isolated incidents. At least two others are known: Archimedes (287 to 212 B.C.) in Ancient Greece and Bhāskara II (A.D. 1114 to 1185) in medieval India developed calculus ideas long before. Tragically, the revolutionary nature of these discoveries either wasn't recognized or else was so buried in other new and difficult-to-understand ideas that they were nearly forgotten until modern times.

The word "calculus" has a modest origin, deriving from similar words such as "calculation" and "calculate," but all these words derive from a Latin (or perhaps even older) root meaning "pebble." In the ancient world, calculi were stone beads used to keep track of livestock and grain reserves (and today, calculi are small stones that form in the gallbladder, kidneys or other parts of the body).

One of the foremost branches of mathematics is calculus. The formal study of calculus started from the 17th century by well-known scientists and mathematicians like Isaac Newton and Gottfried Leibniz, although it is possible that it has been at use as early as the Greek era. It is a mathematical discipline that is primarily concerned with functions, limits, derivatives, and integrals just to name a few. This discipline has a unique legacy over the history of mathematics. Even though it is split between the 2 definitions of Newton and Leibniz, it has still been able to create a new mathematical system and was used in a variety of applications.

There are 2 different fields of calculus. The first subfield is called differential calculus. Using the concept of function derivatives, it studies the behavior and rate on how different quantities change. Using the process of differentiation, the graph of a function can actually be computed, analyzed, and predicted. The second subfield is called integral calculus. Integration is actually the reverse process of differentiation, concerned with the concept of the anti-derivative. Either a concept, or at least semblances of it, has existed for centuries already. Even though these 2 subfields are generally different form each other, these 2 concepts are linked by the fundamental theorem of calculus.

Principles

Limits and Infinitesimals

Calculus is usually developed by working with very small quantities. Historically, the first method of doing so was by infinitesimals. These are objects which can be treated like real numbers but which are, in some sense, "infinitely small". For example, an infinitesimal number could be greater than 0, but less than any number in the sequence 1, 1/2, 1/3, ... and thus less than any positive real number. From this point of view, calculus is a collection of techniques for manipulating infinitesimals. The symbols dx and s dy were taken to be infinitesimal, and the derivative dy / dx was simply their ratio.

The infinitesimal approach fell out of favor in the 19th century because it was difficult to make the notion of an infinitesimal precise. However, the concept was revived in the 20th century with the introduction of non-standard analysis and smooth infinitesimal analysis, which provided solid foundations for the manipulation of infinitesimals.

In the 19th century, infinitesimals were replaced by the epsilon, delta approach to limits. Limits describe the value of a function at a certain input in terms of its values at nearby inputs. They capture small-scale behavior in the context of the real number system. In this treatment, calculus is a collection of techniques for manipulating certain limits. Infinitesimals get replaced by very small numbers, and the infinitely small behavior of the function is found by taking the limiting behavior for smaller and smaller numbers. Limits were the first way to provide rigorous foundations for calculus, and for this reason they are the standard approach.

Leibniz Notation

A common notation, introduced by Leibniz, for the derivative in the example above is

$$y = x^2$$

$$\frac{dy}{dx} = 2x.$$

In an approach based on limits, the symbol $\frac{dy}{dx}$ is to be interpreted not as the quotient of two numbers but as a shorthand for the limit computed above. Leibniz, however, did intend it to represent the quotient of two infinitesimally small numbers, dy being the infinitesimally small change in y caused by an infinitesimally small change dx applied to x. We can also think of $\frac{d}{dx}$ as a differentiation operator, which takes a function as an input and gives another function, the derivative, as the output. For example:

$$\frac{d}{dx}(x^2) = 2x.$$

In this usage, the dx in the denominator is read as "with respect to x". Another example of correct notation could be:

$$g(t) = t^2 + 2t + 4$$

$$\frac{d}{dt} g(t) = 2t + 2$$

Even when calculus is developed using limits rather than infinitesimals, it is common to manipulate symbols like dx and dy as if they were real numbers; although it is possible to avoid such manipulations, they are sometimes notationally convenient in expressing operations such as the total derivative.

Differential Calculus

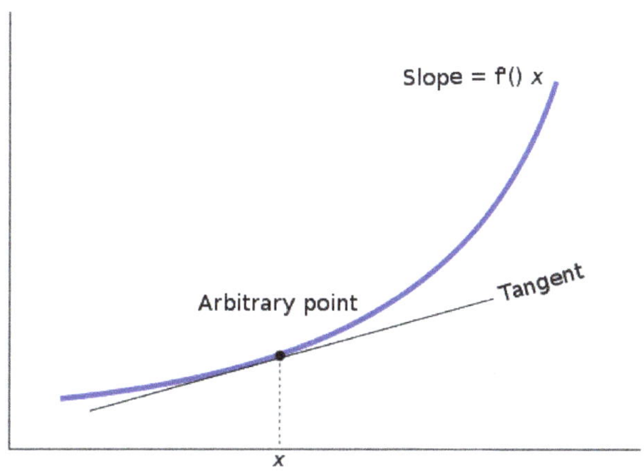

Tangent line at $(x, f(x))$. The derivative $f'(x)$ of a curve at a point is the slope (rise over run) of the line tangent to that curve at that point.

Differential calculus is the study of the definition, properties, and applications of the derivative of a function. The process of finding the derivative is called *differentiation*. Given a function and a point in the domain, the derivative at that point is a way of encoding the small-scale behavior of the function near that point. By finding the derivative of a function at every point in its domain, it is possible to produce a new function, called the *derivative function* or just the *derivative* of the original function. In formal terms, the derivative is a linear operator which takes a function as its input and produces a second function as its output. This is more abstract than many of the processes studied in elementary algebra, where functions usually input a number and output another number. For example, if the doubling function is given the input three, then it outputs six, and if the squaring function is given the input three, then it outputs nine. The derivative, however, can take the squaring function as an input. This means that the derivative takes all the information of the squaring function—such as that two is sent to four, three is sent to nine, four is sent to sixteen, and so on—and uses this information to produce another function. The function produced by deriving the squaring function turns out to be the doubling function.

In more explicit terms the "doubling function" may be denoted by $g(x) = 2x$ and the "squaring function" by $f(x) = x^2$. The "derivative" now takes the function $f(x)$, defined by the expression "x^2", as an input, that is all the information —such as that two is sent to four, three is sent to nine, four is sent to sixteen, and so on— and uses this information to output another function, the function $g(x) = 2x$, as will turn out.

The most common symbol for a derivative is an apostrophe-like mark called prime. Thus, the derivative of a function called f is denoted by f', pronounced "f prime". For instance, if $f(x) = x^2$ is the squaring function, then $f'(x) = 2x$ is its derivative (the doubling function g from above). This notation is known as Lagrange's notation.

If the input of the function represents time, then the derivative represents change with respect to time. For example, if f is a function that takes a time as input and gives the position of a ball at that time as output, then the derivative of f is how the position is changing in time, that is, it is the velocity of the ball.

If a function is linear (that is, if the graph of the function is a straight line), then the function can be written as $y = mx + b$, where x is the independent variable, y is the dependent variable, b is the y-intercept, and:

$$m = \frac{\text{rise}}{\text{run}} = \frac{\text{change in } y}{\text{change in } x} = \frac{\Delta y}{\Delta x}.$$

This gives an exact value for the slope of a straight line. If the graph of the function is not a straight line, however, then the change in y divided by the change in x varies. Derivatives give an exact meaning to the notion of change in output with respect to change in input. To be concrete, let f be a function, and fix a point a in the domain of f. $(a, f(a))$ is a point on the graph of the function. If h is a number close to zero, then $a + h$ is a number close to a. Therefore, $(a + h, f(a + h))$ is close to $(a, f(a))$. The slope between these two points is

$$m = \frac{f(a+h) - f(a)}{(a+h) - a} = \frac{f(a+h) - f(a)}{h}.$$

This expression is called a *difference quotient*. A line through two points on a curve is called a *secant line*, so m is the slope of the secant line between $(a, f(a))$ and $(a + h, f(a + h))$. The secant line is only an approximation to the behavior of the function at the point a because it does not account for what happens between a and $a + h$. It is not possible to discover the behavior at a by setting h to zero because this would require dividing by zero, which is undefined. The derivative is defined by taking the limit as h tends to zero, meaning that it considers the behavior of f for all small values of h and extracts a consistent value for the case when h equals zero:

$$\lim_{h \to 0} \frac{f(a+h) - f(a)}{h}.$$

Geometrically, the derivative is the slope of the tangent line to the graph of f at a. The tangent line is a limit of secant lines just as the derivative is a limit of difference quotients. For this reason, the derivative is sometimes called the slope of the function f.

Here is a particular example, the derivative of the squaring function at the input 3. Let $f(x) = x^2$ be the squaring function.

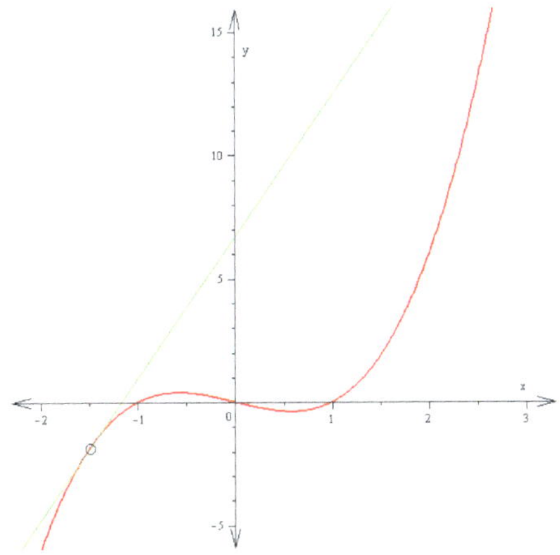

The derivative $f'(x)$ of a curve at a point is the slope of the line tangent to that curve at that point. Here the function involved (drawn in red) is $f(x) = x^3 - x$. The tangent line (in green) which passes through the point (−3/2, −15/8) has a slope of 23/4.

$$\begin{aligned} f'(3) &= \lim_{h\to 0}\frac{(3+h)^2-3^2}{h} \\ &= \lim_{h\to 0}\frac{9+6h+h^2-9}{h} \\ &= \lim_{h\to 0}\frac{6h+h^2}{h} \\ &= \lim_{h\to 0}(6+h) \\ &= 6 \end{aligned}$$

The slope of the tangent line to the squaring function at the point (3, 9) is 6, that is to say, it is going up six times as fast as it is going to the right. The limit process just described can be performed for any point in the domain of the squaring function. This defines the *derivative function* of the squaring function, or just the *derivative* of the squaring function for short. A computation similar to the one above shows that the derivative of the squaring function is the doubling function.

Integral Calculus

Integral calculus is the study of the definitions, properties, and applications of two related concepts, the *indefinite integral* and the *definite integral*. The process of finding the value of an integral is called *integration*. In technical language, integral calculus studies two related linear operators.

The *indefinite integral*, also known as the *antiderivative*, is the inverse operation to the derivative. *F* is an indefinite integral of *f* when *f* is a derivative of *F*. (This use of lower- and upper-case letters for a function and its indefinite integral is common in calculus.)

The *definite integral* inputs a function and outputs a number, which gives the algebraic sum of areas between the graph of the input and the x-axis. The technical definition of the definite integral involves the limit of a sum of areas of rectangles, called a Riemann sum.

A motivating example is the distances traveled in a given time.

Distance = Speed•Time

If the speed is constant, only multiplication is needed, but if the speed changes, a more powerful method of finding the distance is necessary. One such method is to approximate the distance traveled by breaking up the time into many short intervals of time, then multiplying the time elapsed in each interval by one of the speeds in that interval, and then taking the sum (a Riemann sum) of the approximate distance traveled in each interval. The basic idea is that if only a short time elapses, then the speed will stay more or less the same. However, a Riemann sum only gives an approximation of the distance traveled. We must take the limit of all such Riemann sums to find the exact distance traveled.

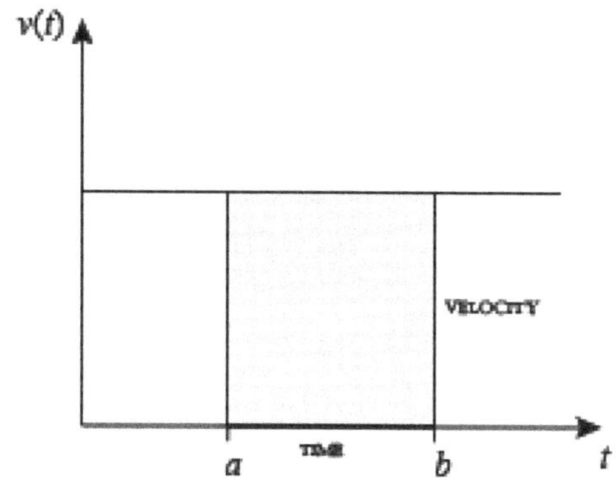

Constant velocity

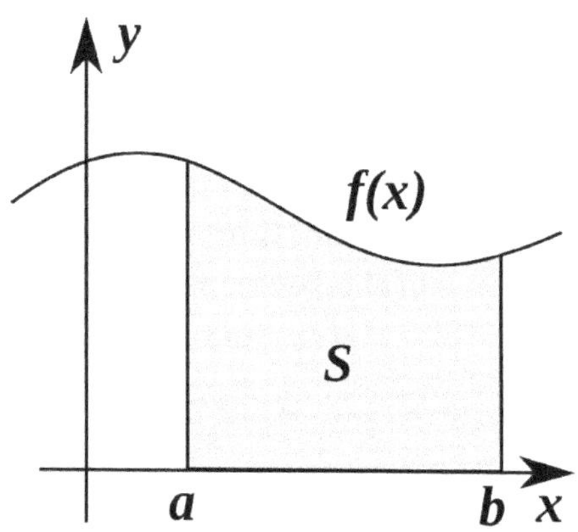

Integration can be thought of as measuring the area under a curve, defined by $f(x)$, between two points (here a and b).

When velocity is constant, the total distance traveled over the given time interval can be computed by multiplying velocity and time. For example, travelling a steady 50 mph for 3 hours results in a total distance of 150 miles. In the diagram, when constant velocity and time are graphed, these two values form a rectangle with height equal to the velocity and width equal to the time elapsed. Therefore, the product of velocity and time also calculates the rectangular area under the (constant) velocity curve. This connection between the area under a curve and distance traveled can be extended to *any* irregularly shaped region exhibiting a fluctuating velocity over a given time period. If $f(x)$ in the diagram represents speed as it varies over time, the distance traveled (between the times represented by a and b) is the area of the shaded region s.

To approximate that area, an intuitive method would be to divide up the distance between a and b into a number of equal segments, the length of each segment represented by the symbol Δx. For each small segment, we can choose one value of the function $f(x)$. Call that value h. Then the area of the rectangle with base Δx and height h gives the distance (time Δx multiplied by speed h) traveled in that segment. Associated with each segment is the average value of the function above it, $f(x) = h$. The sum of all such rectangles gives an approximation of the area between the axis and the curve, which is an approximation of the total distance traveled. A smaller value for Δx will give more rectangles and in most cases a better approximation, but for an exact answer we need to take a limit as Δx approaches zero.

The symbol of integration is $\int$, an elongated S (the S stands for "sum"). The definite integral is written as:

$$\int_a^b f(x)dx$$

and is read "the integral from a to b of f-of-x with respect to x." The Leibniz notation dx is intended to suggest dividing the area under the curve into an infinite number of rectangles, so that their width Δx becomes the infinitesimally small dx. In a formulation of the calculus based on limits, the notation

$$\int_a^b \cdots dx$$

is to be understood as an operator that takes a function as an input and gives a number, the area, as an output. The terminating differential, dx, is not a number, and is not being multiplied by $f(x)$, although, serving as a reminder of the Δx limit definition, it can be treated as such in symbolic manipulations of the integral. Formally, the differential indicates the variable over which the function is integrated and serves as a closing bracket for the integration operator.

The indefinite integral, or antiderivative, is written

$$\int f(x)dx.$$

Functions differing by only a constant have the same derivative, and it can be shown that the antiderivative of a given function is actually a family of functions differing only by a constant. Since the derivative of the function $y = x^2 + C$, where C is any constant, is $y' = 2x$, the antiderivative of the latter given by:

$$\int 2x\, dx = x^2 + C.$$

The unspecified constant C present in the indefinite integral or antiderivative is known as the constant of integration.

Fundamental Theorem

The fundamental theorem of calculus states that differentiation and integration are inverse operations. More precisely, it relates the values of antiderivatives to definite integrals. Because it is usually easier to compute an antiderivative than to apply the definition of a definite integral, the fundamental theorem of calculus provides a practical way of computing definite integrals. It can also be interpreted as a precise statement of the fact that differentiation is the inverse of integration.

The fundamental theorem of calculus states: If a function f is continuous on the interval $[a, b]$ and if F is a function whose derivative is f on the interval (a, b), then

$$\int_a^b f(x)dx = F(b) - F(a).$$

Furthermore, for every x in the interval (a, b),

$$\frac{d}{dx}\int_a^x f(t)dt = f(x).$$

This realization, made by both Newton and Leibniz, who based their results on earlier work by Isaac Barrow, was key to the proliferation of analytic results after their work became known. The fundamental theorem provides an algebraic method of computing many definite integrals—without performing limit processes—by finding formulas for antiderivatives. It is also a prototype solution of a differential equation. Differential equations relate an unknown function to its derivatives, and are ubiquitous in the sciences.

Varieties

Over the years, many reformulations of calculus have been investigated for different purposes.

Non-standard Calculus

Imprecise calculations with infinitesimals were widely replaced with the rigorous (ϵ, δ)-definition of limit starting in the 1870s. Meanwhile, calculations with infinitesimals persisted and often led to correct results. This led Abraham Robinson to investigate if it were possible to develop a number system with infinitesimal quantities over which the theorems of calculus were still valid. In 1960, building upon the work of Edwin Hewitt and Jerzy Łoś, he succeeded in developing non-standard analysis. The theory of non-standard analysis is rich enough to be applied in many branches of mathematics. As such, books and articles dedicated solely to the traditional theorems of calculus often go by the title non-standard calculus.

Smooth Infinitesimal Analysis

This is another reformulation of the calculus in terms of infinitesimals. Based on the ideas of F. W. Lawvere and employing the methods of category theory, it views all functions as being continuous and incapable of being expressed in terms of discrete entities. One aspect of this formulation is that the law of excluded middle does not hold in this formulation.

Constructive Analysis

Constructive mathematics is a branch of mathematics that insists that proofs of the existence of a number, function, or other mathematical object should give a construction of the object. As such constructive mathematics also rejects the law of excluded middle. Reformulations of calculus in a constructive framework are generally part of the subject of constructive analysis.

Applications

Calculus is used in every branch of the physical sciences, actuarial science, computer science, statistics, engineering, economics, business, medicine, demography, and in other fields wherever a problem can be mathematically modeled and an optimal solution is desired. It allows one to go from (non-constant) rates of change to the total change or vice versa, and many times in studying a problem we know one and are trying to find the other.

The logarithmic spiral of the Nautilus shell is a classical image used to depict the growth and change related to calculus.

Physics makes particular use of calculus; all concepts in classical mechanics and electromagnetism are related through calculus. The mass of an object of known density, the moment of inertia of objects, as well as the total energy of an object within a conservative field can be found by the use of calculus. An example of the use of calculus in mechanics is Newton's second law of motion: historically stated it expressly uses the term "change of motion" which implies the derivative saying The change of momentum of a body is equal to the resultant force acting on the body and is in the same direction. Commonly expressed today as Force = Mass × acceleration, it implies differential calculus because acceleration is the time derivative of velocity or second time derivative of trajectory or spatial position. Starting from knowing how an object is accelerating, we use calculus to derive its path.

Maxwell's theory of electromagnetism and Einstein's theory of general relativity are also expressed in the language of differential calculus. Chemistry also uses calculus in determining reaction rates and radioactive decay. In biology, population dynamics starts with reproduction and death rates to model population changes.

Calculus can be used in conjunction with other mathematical disciplines. For example, it can be used with linear algebra to find the "best fit" linear approximation for a set of points in a domain. Or it can be used in probability theory to determine the probability of a continuous random variable from an assumed density function. In analytic geometry, the study of graphs of functions, calculus is used to find high points and low points (maxima and minima), slope, concavity and inflection points.

Green's Theorem, which gives the relationship between a line integral around a simple closed curve C and a double integral over the plane region D bounded by C, is applied in an instrument known as a planimeter, which is used to calculate the area of a flat surface on a drawing. For example, it can be used to calculate the amount of area taken up by an irregularly shaped flower bed or swimming pool when designing the layout of a piece of property.

Discrete Green's Theorem, which gives the relationship between a double integral of a function around a simple closed rectangular curve C and a linear combination of the antiderivative's values at corner points along the edge of the curve, allows fast calculation of sums of values in rectangular domains. For example, it can be used to efficiently calculate sums of rectangular domains in images, in order to rapidly extract features and detect object; another algorithm that could be used is the summed area table.

In the realm of medicine, calculus can be used to find the optimal branching angle of a blood vessel so as to maximize flow. From the decay laws for a particular drug's elimination from the body, it is used to derive dosing laws. In nuclear medicine, it is used to build models of radiation transport in targeted tumor therapies.

In economics, calculus allows for the determination of maximal profit by providing a way to easily calculate both marginal cost and marginal revenue.

Calculus is also used to find approximate solutions to equations; in practice it is the standard way to solve differential equations and do root finding in most applications. Examples are methods such as Newton's method, fixed point iteration, and linear approximation. For instance, spacecraft use a variation of the Euler method to approximate curved courses within zero gravity environments.

Applications of Calculus

Application in Engineering

1. An Architect Engineer uses integration in determining the amount of the necessary materials to construct curved shape constructions (e.g. dome over a sports arena) and also to measure the weight of that structure. Calculus is used to improve the architecture not only of buildings but also of important infrastructures such as bridges.
2. In Electrical Engineering, Calculus (Integration) is used to determine the exact length of power cable needed to connect two substations, which are miles away from each other.
3. Space flight engineers frequently use calculus when planning for long missions. To launch an exploratory probe, they must consider the different orbiting velocities of the Earth and the planet the probe is targeted for, as well as other gravitational influences like the sun and the moon.

Application in Medical Science

1. Biologists use differential calculus to determine the exact rate of growth in a bacterial culture when different variables such as temperature and food source are changed.

Application in Physics

1. In Physics, Integration is very much needed. For example, to calculate the Centre of Mass, Centre of Gravity and Mass Moment of Inertia of a sports utility vehicle.
2. To calculate the velocity and trajectory of an object, predict the position of planets, and understand electromagnetism.

Application in Statistics

1. Statisticians use calculus to evaluate survey data to help develop business plans for different companies. Because a survey involves many different questions with a range of possible answers, calculus allows a more accurate prediction for the appropriate action.

Application in Research Analysis

1. An operations research analyst will use calculus when observing different processes at a manufacturing corporation. By considering the value of different variables, they can help a company improve operating efficiency, increase production, and raise profits.

Application in Graphics

1. A graphics artist uses calculus to determine how different three-dimensional models will behave when subjected to rapidly changing conditions. It can create a realistic environment for movies or video games.

Application in Chemistry

1. It is used to determine the rate of a chemical reaction and to determine some necessary information of Radioactive decay reaction.

Let's Take a Look at Some other Examples

- Credit card companies use calculus to set the minimum payments due on credit card statements at the exact time the statement is processed by considering multiple variables such as changing interest rates and a fluctuating available balance.
- Doctors and lawyers use calculus to help build the discipline necessary for solving complex problems, such as diagnosing patients or planning a prosecution case.
- The field of epidemiology — the study of the spread of infectious disease — relies heavily on calculus. It can be used to determine how far and fast a disease is spreading, where it may have originated from and how to best treat it.

Geometric Calculus

Geometric calculus is an extension of *geometric algebra* to include the analytic operations of differentiation and integration. It is developed in this book as a computational language for physics. The calculus is designed for efficient computation and representation of geometric relations. This leads to compact formulations for the equations of physics and their solutions as well as elucidation of their geometric contents.

The gradient, or vector derivative, is $\nabla = \sum_j e_j \partial_j$, where ∂_j is a partial derivative. The gradient acts algebraically as a vector when applied to scalar or vector valued functions: we can multiply it by a scalar field f, giving the vector field ∇f; dot it with a vector field f, giving the scalar field ∇.f; and in $\mathbb{R}^3$ cross it with f, giving the vector field $\nabla \times \mathrm{f}$ But $\nabla \mathrm{f}$, a product of vectors, cannot be formed in vector calculus.

In geometric calculus $\nabla \mathrm{f}$ does make sense. This product of vectors is scalar + bivector: $\nabla \mathrm{f} = \nabla.\mathrm{f} + \nabla \wedge \mathrm{f}$, just as with the fundamental identity, Equation. In this way ∇ unifies the divergence and curl (generalized to $\wedge$)

- The geometric calculus identity $\nabla^2 \mathrm{f} = \nabla(\nabla \mathrm{f})$ cannot be written in vector calculus. Instead, we must resort to $\nabla^2 \mathrm{f} = \nabla(\nabla \cdot \mathrm{f}) - \nabla \times (\nabla \times \mathrm{f})$, and this only in $\mathbb{R}^3$.

Electromagnetism. Classical (i.e., non-quantum) electromagnetic theory is usually formulated in $\mathbb{R}^3$ vector calculus. Two vector fields, the electric field e and the magnetic field b, represent the electromagnetic field. Maxwell's equations are the heart of the theory. In a vacuum they are:

$$\nabla \bullet \mathrm{e} = 0, \nabla \times \mathrm{b} = \partial_t \mathrm{e}, \quad \nabla \times \mathrm{e} = -\partial_t \mathrm{b}, \quad \nabla \bullet \mathrm{b} = 0.$$

The geometric calculus formulation of Maxwell's theory represents a magnetic field with the bivector $\mathrm{B} = -\mathrm{b}^*$. Maxwell's equations become, using the duality relations Equation and $(\mathrm{u} \wedge \mathrm{v})^* = \mathrm{u} \times \mathrm{v}$.

$$\nabla \bullet \mathrm{e} = 0, \nabla \bullet \mathrm{B} = -\partial_t \mathrm{e}, \ \nabla \wedge \mathrm{e} = -\partial_t \mathrm{B}, \ \nabla \wedge \mathrm{B} = 0.$$

The geometric calculus formulation represents the electromagnetic field with the single multivector field $F = e + B$, vector + bivector. Maxwell's equations become $(\partial_t + \nabla)F = 0$; the scalar, vector, bivector, trivector parts of this equation are precisely Equation, as you can check.

The electromagnetic field obeys the wave equation:

$$\left(\partial^2{}_t - \nabla^2\right)F = (\partial_t - \nabla)(\partial_t + \nabla)F = 0.$$

- The vector calculus derivation of this result is not straightforward.
- Calculations using the G^3 formulation of Maxwell's equations are often easier, and sometimes much easier, than using the $\mathbb{R}^3$ formulation. This is due in part to the fact that the GA derivative ∇, unlike the divergence ∇.and curl $\nabla \times$ in Equation, is invertible.

Analytic functions. Let $f(x, y) = u(x, y) + v(x, y)\mathrm{i}$, with u and v real valued. Then

$$\nabla f = \mathrm{e}_1 \left(u_x + v_x \mathrm{i}\right) + \mathrm{e}_2 \left(u_y + v_y \mathrm{i}\right) = \mathrm{e}_1 \left(u_x - v_y\right) + \mathrm{e}_2 \left(v_x + u_y\right).$$

From the Cauchy-Riemann equations, $\nabla f = 0 \Leftrightarrow f$ is analytic. Generalizing, we call a multivector field F on $\mathbb{R}^n$ analytic if $\nabla F = 0$.

- This definition leads to a generalization of standard complex analysis to n (real) dimensions. Many standard results generalize. A simple example: since $\nabla F = 0 \Rightarrow \nabla^2 F = \nabla(\nabla F) = 0$, analytic functions are harmonic functions. Most important, Cauchy's theorem and Cauchy's integral formula generalize.
- There is no multivector field called the derivative of F.

Generalize ∇. It violates the spirit of GA to write $f(x,y)$ above as a function of coordinates. Henceforth we shall think of it, equivalently, as a function of the vector $\mathrm{x} = \mathrm{e}_1\mathrm{x} + \mathrm{e}_2\mathrm{y}$, and similarly in higher dimensions. For example, for a constant multivector $M, \nabla(\mathrm{x}\, M) = nM,$ as you can verify. With this change of viewpoint we can generalize ∇.First, the directional derivative of F in the "direction" A is

$$\partial_A F(X) = \lim_{\tau \to 0} \frac{F(X + \tau P(A)) - F(X)}{\tau},$$

where p_x projects onto the grades of X. The projection ensures, for exam- ple, that the directional derivative of a function F of vectors only has a zero directional derivative in the directions other than vectors.

Abbreviating ∂_{eJ} to $\partial_{J,}$ we generalize the vector derivative $\nabla = \sum_j e_j \partial_j$ to the multivector derivative:

$$\nabla = \sum_J e_j^{-1} \partial_j.$$

This is a Fourier expansion of ∇. It can operate on any multivector valued function.

Minimization example. As an application of the multivector derivative, consider the problem of rotating 3D vectors u_1; u_2....., u_n to best approximate the vectors v_1; v_2;...., v_n. By this we mean choose a rotor R to minimize $\phi(R) = \sum_j \left(v_j - Ru_j R^{-1}\right)^2$. We show that the minimum occurs when $\nabla\phi(R) = 0$.

First note that we may drop the constraint $|R| = 1,$ since $\phi(aR)$ $\phi(R)$ Set $R = r_0 + e_3 e_2 r_1 + e_1 e_3 r_2 + e_2 e_1 r_3$. Then

$$\phi(R) = \phi\left(r_0 + e_3 e_2 r_1 + e_1 e_3 r_2 + e_2 e_1 r_3\right)$$

is a real valued function of four real variables. Its minimum occurs when all four partial derivatives $\partial\phi / \partial r_i = 0$. We have

$$\nabla\phi(R) = 1\partial\phi / \partial r_0 + e_2 e_3 \partial\phi / \partial r_1 + e_3 e_1 \partial\phi / \partial r_2 + e_1 e_2 \partial\phi / \partial r_3.$$

Thus the minimum occurs when $\nabla\phi(R) = 0$.

After learning a few rules about differentiating with respect to a multivector we find that $\nabla\phi(R) = 0 \Rightarrow \sum_j v_j \wedge \left(Ru_j R^{-1}\right) = 0$. The right side has an agreeable geometric interpretation: the bivector sum of the parallelograms spanned by the v_j and the rotated u_j is zero. The equation can be solved for R.

- Geometric calculus uses only one derivative, $\nabla\phi$, to solve this problem. Vector calculus uses the four components $\partial\phi / \partial r_i$ of $\nabla\phi$ It is generally best not to break a multivector into its components, just as it is generally best not to break a complex number into its real and imaginary parts or a vector into its components.

Differentiation

With a geometric algebra given, let *a* and *b* be vectors and let *F(a)* be a multivector-valued function. The directional derivative of *F(a)* along *b* is defined as

$$\nabla_b F(a) = \lim_{\epsilon \to 0} \frac{F(a+\epsilon b) - F(a)}{\epsilon}$$

provided that the limit exists, where the limit is taken for scalar ϵ. This is similar to the usual definition of a directional derivative but extends it to functions that are not necessarily scalar-valued.

Next, choose a set of basis vectors $\{e_i\}$ and consider the operators, noted (∂_i), that perform directional derivatives in the directions of (e_i):

$$\partial_i : F \mapsto (x \mapsto \nabla_{e_i} F(x))$$

Then, using the Einstein summation notation, consider the operator :

$$e^i \partial_i$$

which means:

$$F \mapsto e^i \partial_i F$$

or, more verbosely:

$$F \mapsto (x \mapsto e^i \nabla_{e_i} F(x))$$

It can be shown that this operator is independent of the choice of frame, and can thus be used to define the *geometric derivative*:

$$\nabla = e^i \partial_i$$

This is similar to the usual definition of the gradient, but it, too, extends to functions that are not necessarily scalar-valued.

It can be shown that the directional derivative is linear regarding its direction, that is:

$$\nabla_{\alpha a + \beta b} = \alpha \nabla_a + \beta \nabla_b$$

From this follows that the directional derivative is the inner product of its direction by the geometric derivative. All needs to be observed is that the direction a can be written $a=(a\cdot e^i)e_i$, so that:

$$\nabla_a=\nabla_{(a\cdot e^i)e_i}=(a\cdot e^i)\nabla_{e_i}=a\cdot(e^i\nabla_{e^i})=a\cdot\nabla$$

For this reason, $\nabla_a F(x)$ is often noted $a\cdot\nabla F(x)$.

The standard order of operations for the geometric derivative is that it acts only on the function closest to its immediate right. Given two functions *F* and *G*, then for example we have

$$\nabla FG=(\nabla F)G.$$

Product Rule

Although the partial derivative exhibits a product rule, the geometric derivative only partially inherits this property. Consider two functions *F* and *G*:

$$\begin{aligned}\nabla(FG)&=e^i\partial_i(FG)\\&=e^i((\partial_i F)G+F(\partial_i G))\\&=e^i(\partial_i F)G+e^iF(\partial_i G)\end{aligned}$$

Since the geometric product is not commutative with $e^iF\neq Fe^i$ in general, we cannot proceed further without new notation. A solution is to adopt the *overdot notation*, in which the scope of a geometric derivative with an overdot is the multivector-valued function sharing the same overdot. In this case, if we define

$$\dot\nabla F\dot G=e^iF(\partial_i G),$$

then the product rule for the geometric derivative is

$$\nabla(FG)=\nabla FG+\dot\nabla F\dot G$$

Interior and Exterior Derivative

Let *F* be an *r*-grade multivector. Then we can define an additional pair of operators, the interior and exterior derivatives,

$$\nabla\cdot F=\langle\nabla F\rangle_{r-1}=e^i\cdot\partial_i F$$

$$\nabla\wedge F=\langle\nabla F\rangle_{r+1}=e^i\wedge\partial_i F.$$

In particular, if *F* is grade 1 (vector-valued function), then we can write

$$\nabla F=\nabla\cdot F+\nabla\wedge F$$

and identify the divergence and curl as

$$\nabla \cdot F = \operatorname{div} F$$

$$\nabla \wedge F = I \operatorname{curl} F.$$

However, that these two operators are considerably weaker than the geometric derivative counterpart for several reasons. Neither the interior derivative operator nor the exterior derivative operator is invertible.

Integration

Let $\{e_1,...e_n\}$ be a set of basis vectors that span an n-dimensional vector space. From geometric algebra, we interpret the pseudoscalar $e_1 \wedge e_2 \wedge \cdots \wedge e_n$ to be the signed volume of the n-parallelotope subtended by these basis vectors. If the basis vectors are orthonormal, then this is the unit pseudoscalar.

More generally, we may restrict ourselves to a subset of k of the basis vectors, where $1 \leq k \leq n$, to treat the length, area, or other general k-volume of a subspace in the overall n-dimensional vector space. We denote these selected basis vectors by $\{e_{i_1},...e_{i_k}\}$. A general k-volume of the k-parallelotope subtended by these basis vectors is the grade k multivector $e_{i_1} \wedge e_{i_2} \wedge \cdots \wedge e_{i_k}$.

Even more generally, we may consider a new set of vectors $\{x^{i_1}e_{i_1},...x^{i_k}e_{i_k}\}$ proportional to the k basis vectors, where each of the $\{x^{i_j}\}$ is a component that scales one of the basis vectors. We are free to choose components as infinitesimally small as we wish as long as they remain nonzero. Since the outer product of these terms can be interpreted as a k-volume, a natural way to define a measure is

$$\begin{aligned} d^k X &= \left(dx^{i_1}e_{i_1}\right) \wedge \left(dx^{i_2}e_{i_2}\right) \wedge \cdots \wedge \left(dx^{i_k}e_{i_k}\right) \\ &= \left(e_{i_1} \wedge e_{i_2} \wedge \cdots \wedge e_{i_k}\right) dx^{i_1} dx^{i_2} \cdots dx^{i_k} \end{aligned}$$

The measure is therefore always proportional to the unit pseudoscalar of a k-dimensional subspace of the vector space. Compare the Riemannian volume form in the theory of differential forms. The integral is taken with respect to this measure:

$$\int_V F(x) d^k X = \int_V F(x) \left(e_{i_1} \wedge e_{i_2} \wedge \cdots \wedge e_{i_k}\right) dx^{i_1} dx^{i_2} \cdots dx^{i_k}$$

More formally, consider some directed volume V of the subspace. We may divide this volume into a sum of simplices. Let $\{x_i\}$ be the coordinates of the vertices. At each vertex we assign a measure $\Delta U_i(x)$ as the average measure of the simplices sharing the vertex. Then the integral of *F(x)* with respect to *U(x)* over this volume is obtained in the limit of finer partitioning of the volume into smaller simplices:

$$\int_V F dU = \lim_{n \to \infty} \sum_{i=1}^{n} F(x_i) \Delta U_i(x).$$

Fundamental Theorem of Geometric Calculus

The reason for defining the geometric derivative and integral as above is that they allow a strong generalization of Stokes' theorem. Let $\mathsf{L}(A;x)$ be a multivector-valued function of r-grade input A and general position x, linear in its first argument. Then the fundamental theorem of geometric calculus relates the integral of a derivative over the volume V to the integral over its boundary:

$$\int_V \dot{\mathsf{L}}\left(\dot{\nabla} dX; x\right) = \oint_{\partial V} \mathsf{L}(dS; x)$$

As an example, let $\mathsf{L}(A;x) = \langle F(x) A I^{-1} \rangle$ for a vector-valued function $F(x)$ and a (n-1)-grade multivector A. We find that

$$\begin{aligned}\int_V \dot{\mathsf{L}}\left(\dot{\nabla} dX; x\right) &= \int_V \langle \dot{F}(x) \dot{\nabla} dX I^{-1} \rangle \\ &= \int_V \langle \dot{F}(x) \dot{\nabla} \mid dX \mid \rangle \\ &= \int_V \nabla \cdot F(x) \mid dX \mid .\end{aligned}$$

and likewise

$$\begin{aligned}\oint_{\partial V} \mathsf{L}(dS; x) &= \oint_{\partial V} \langle F(x) dS I^{-1} \rangle \\ &= \oint_{\partial V} \langle F(x) \hat{n} \mid dS \mid \rangle \\ &= \oint_{\partial V} F(x) \cdot \hat{n} \mid dS \mid\end{aligned}$$

Thus we recover the divergence theorem,

$$\int_V \nabla \cdot F(x) \mid dX \mid = \oint_{\partial V} F(x) \cdot \hat{n} \mid dS \mid .$$

Covariant Derivative

A sufficiently smooth k-surface in an n-dimensional space is deemed a manifold. To each point on the manifold, we may attach a k-blade B that is tangent to the manifold. Locally, B acts as a pseudoscalar of the k-dimensional space. This blade defines a projection of vectors onto the manifold:

$$\mathcal{P}_B(A) = (A \cdot B^{-1}) B$$

Just as the geometric derivative ∇ is defined over the entire n-dimensional space, we may wish to define an *intrinsic derivative* ∂, locally defined on the manifold:

$$\partial F = \mathcal{P}_B(\nabla) F$$

(The right hand side of the above may not lie in the tangent space to the manifold. Therefore, it is not the same as $\mathcal{P}_B(\nabla F)$, which necessarily does lie in the tangent space.)

If a is a vector tangent to the manifold, then indeed both the geometric derivative and intrinsic derivative give the same directional derivative

$$a \cdot \partial F = a \cdot \nabla F$$

Although this operation is perfectly valid, it is not always useful because ∂F itself is not necessarily on the manifold. Therefore, we define the covariant derivative to be the forced projection of the intrinsic derivative back onto the manifold:

$$a \cdot DF = \mathcal{P}_B(a \cdot \partial F) = \mathcal{P}_B(a \cdot \mathcal{P}_B(\nabla F))$$

Since any general multivector can be expressed as a sum of a projection and a rejection, in this case

$$a \cdot \partial F = \mathcal{P}_B(a \cdot \partial F) + \mathcal{P}_B^{\perp}(a \cdot \partial F),$$

we introduce a new function, the shape tensor $\mathsf{S}(a)$, which satisfies

$$F \times \mathsf{S}(a) = \mathcal{P}_B^{\perp}(a \cdot \partial F),$$

where $\times$ is the commutator product. In a local coordinate basis $\{e_i\}$ spanning the tangent surface, the shape tensor is given by

$$\mathsf{S}(a) = e^i \wedge \mathcal{P}_B^{\perp}(a \cdot \partial e_i).$$

Importantly, on a general manifold, the covariant derivative does not commute. In particular, the commutator is related to the shape tensor by

$$[a \cdot D, b \cdot D]F = -(\mathsf{S}(a) \times \mathsf{S}(b)) \times F.$$

Clearly the term $\mathsf{S}(a) \times \mathsf{S}(b)$ is of interest. However it, like the intrinsic derivative, is not necessarily on the manifold. Therefore, we can define the Riemann tensor to be the projection back onto the manifold:

$$\mathsf{R}(a \wedge b) = -\mathcal{P}_B(\mathsf{S}(a) \times \mathsf{S}(b)).$$

Lastly, if F is of grade r, then we can define interior and exterior covariant derivatives as

$$D \cdot F = \langle DF \rangle_{r-1}$$

$$D \wedge F = \langle DF \rangle_{r+1},$$

and likewise for the intrinsic derivative.

Relation to Differential Geometry

On a manifold, locally we may assign a tangent surface spanned by a set of basis vectors $\{e_i\}$. We can associate the components of a metric tensor, the Christoffel symbols, and the Riemann tensor as follows:

$$g_{ij} = e_i \cdot e_j$$

$$\Gamma^k_{ij} = (e_i \cdot De_j) \cdot e^k$$

$$R_{ijkl} = (\mathsf{R}(e_i \wedge e_j) \cdot e_k) \cdot e_l$$

These relations embed the theory of differential geometry within geometric calculus.

Relation to Differential Forms

In a local coordinate system $(x^1, ..., x^n)$, the coordinate differentials $dx^1, ..., dx^n$ form a basic set of one-forms within the coordinate chart. Given a multi-index $i_1, ..., i_k$ with $1 \le i_p \le n$ for $1 \le p \le k$, we can define a k-form

$$\omega = f_I dx^I = f_{i_1, i_2 \cdots i_k} dx^{i_1} \wedge dx^{i_2} \wedge \cdots \wedge dx^{i_k}.$$

We can alternatively introduce a k-grade multivector A as

$$A = f_{i_1, i_2 \cdots i_k} e^{i_1} \wedge e^{i_2} \wedge \cdots \wedge e^{i_k}$$

and a measure

$$\begin{aligned} d^k X &= \left(dx^{i_1} e_{i_1}\right) \wedge \left(dx^{i_2} e_{i_2}\right) \wedge \cdots \wedge \left(dx^{i_k} e_{i_k}\right) \\ &= \left(e_{i_1} \wedge e_{i_2} \wedge \cdots \wedge e_{i_k}\right) dx^{i_1} dx^{i_2} \cdots dx^{i_k} \end{aligned}.$$

Apart from a subtle difference in meaning for the exterior product with respect to differential forms versus the exterior product with respect to vectors (indeed one should note that in the former the *increments* are covectors, whereas in the latter they represent scalars), we see the correspondences of the differential form

$$\omega \cong A^\dagger \cdot d^k X = A \cdot \left(d^k X\right)^\dagger,$$

its derivative

$$d\omega \cong (D \wedge A)^\dagger \cdot d^{k+1} X = (D \wedge A) \cdot \left(d^{k+1} X\right)^\dagger,$$

and its Hodge dual

$$\star\omega \cong (I^{-1} A)^\dagger \cdot d^k X,$$

embed the theory of differential forms within geometric calculus.

Limit of a Function

Consider the function f(x). Let the independent variable x take values near a given constant 'a'. Then, f(x) takes a corresponding set of values. Suppose that when x is close to 'a', the values of f(x) are close to some constant. Suppose f(x) can be made to differ arbitrarily small from A by taking values of x that are sufficiently close to 'a' but not equal to 'a' and that is true for all such values of x. Then, f(x) is said to approach limit A as x approaches 'a'.

A function may approach two different limits. One where the variable approaches its limit through values larger than the limit and the other where the variable approaches its limit through values smaller than the limit. In such a case, the limit is not defined but the right and left hand limit exists.

The right hand limit of a function is the value of the function approaches when the variable approaches its limit from the right. This can be written as $\lim_{x \to a} f(x) = A^+$

The left hand limit of a function is the value of the function approaches when the variable approaches its limit from the left. This can be written as $\lim_{x \to a} f(x) = A^-$

The limit of a function exists if and only if the left hand limit is equal to the right hand limit. In that case, $\lim_{x \to a^{-1}} f(x) = \lim_{x \to a^{+1}} f(x) = L$

Right Handed and Left handed limits

A function is said to possess a right hand limit as x approaches a from the values higher than and is expressed as $\lim_{x \to a^+} f(x)$

A function is said to possess a left hand limit as x approaches a from values lower than a, and is expressed as $\lim_{x \to a^-} f(x)$

A function will have a limiting value only if its right hand limit equals its left hand limit.

For example, let f(x) = [x] , where [x] means the greatest integer less than or equal to x.

Let $x \to 2^+$

The values of [x] for values of x greater than 2 and close to 2 are tabulated below.

x	2.1	2.001	2.0001	2,00001
F(x)=[x]	2	2	2	2

Therefore, $\lim_{x \to 2^+} [x] = 2$

Let $x \to 2$

Values of f(x) at points less than 2 and close to 2 are tabulated below.

x	1-9	1.99	1.999	1.9999
f(x)=[x]	1	1	1	1

$\lim_{x \to 2^-} [x] = 1$

Since the left hand limit and the right hand limit are not equal, therefore $\lim_{x \to 2^-} [x]$ does not exist.

Laws of Limits

There are a number of basic laws of limits, by which we can find most of the limits. We can apply them for all forms of limits.

- For a constant function C, we define limit as $\lim_{x \to a} c = c$
- For function f(x) = x, we define limit as $\lim_{x \to a} x = a$

Given below are some of the basic laws of limits:

Law of Addition:

If limits $\lim_{x \to a} f(x)$ and $\lim_{x \to a} g(x)$ exists, then,

$$\lim_{x \to a} (f(x) + g(x)) = \lim_{x \to a} f(x) + \lim_{x \to a} g(x)$$

Law of Subtraction:

If limits $\lim_{x \to a} f(x)$ and $\lim_{x \to a} g(x)$ exists, then,

$$\lim_{x \to a} (f(x) - g(x)) = \lim_{x \to a} f(x) - \lim_{x \to a} g(x)$$

Law of Multiplication:

If limits $\lim_{x \to a} f(x)$ and $\lim_{x \to a} g(x)$ exists, then

$$\lim_{x \to a} (f(x) \times g(x)) = \lim_{x \to a} f(x) \times \lim_{x \to a} g(x)$$

Law of Division:

If limits $\lim_{x \to a} f(x)$ and $\lim_{x \to a} g(x)$ exists, then

$$\lim_{x \to a} \left(\frac{f(x)}{g(x)} \right) = \frac{\lim_{x \to a} f(x)}{\lim_{x \to a} g(x)}$$

Law of Constant:

If limits $\lim_{x\to a} f(x)$ exists, then,

$$\lim_{x\to a}\left(cf(x)\right)=c\lim_{x\to a} f(x)$$

Here, C is a constant.

Law of Power:

If limits $\lim_{x\to a} f(x)$ exists, then,

$$\lim_{x\to a}\left(f(x)\right)^n=(\lim_{x\to a} f(x))^n$$

Here, n is an integer

Law of Root:

If limits $\lim_{x\to a} f(x)$ exists, then,

$$\lim_{x\to a}\sqrt[n]{f(x)}=\sqrt[n]{\lim_{x\to a} f(x)}$$

Here, n is an integer

Certain Special Limits

$1.\lim_{x\to a}\dfrac{x^n-a^n}{x-a}=n\,a(n-1),$ for all the real vales of n.

$2.\lim_{\theta\to 0}\dfrac{\sin\theta}{\theta}=1$

$3.\lim_{\theta\to 0}\dfrac{\tan\theta}{\theta}=1$

$4.\lim_{\cos\to 0}\cos\theta=1$

$5.\lim_{n\to\infty}\left(1+\dfrac{1}{n}\right)^n=e$

$6.\lim_{n\to 0}(1+n)^{\frac{1}{n}}=e$

$7.\lim_{n\to 0}\dfrac{e^x-1}{x}=1$

Examples on Laws of Limits

Given below are some of the examples on laws of limits.

Question 1: Find $\lim_{x\to a} x^n$

Solution:

Step 1: By law of power,

$\lim_{x\to a}(f(x))^n = (\lim_{x\to a} f(x))^n$

Step 2: $\lim_{x\to a}(x)^n = \left(\lim_{x\to a} x\right)^n$

Step 3: $\lim_{x\to a}(x)^n = a^n$

$\lim_{x\to a} x^n = a^n$

Question 2: Find $\lim_{x\to a}\left(cx+dx\right)$ where, c and d are constant

Solution:

Step 1: By law of addition, $\lim_{x\to a}(f(x)+g(x)) = \lim_{x\to a} f(x) + \lim_{x\to a} g(x)$

Step 2: $\lim_{x\to a}\left(cx+dx\right) = \lim_{x\to a} cx + \lim_{x\to a} dx$

Step 3: By Law of constant, $\lim_{x\to a} cx + \lim_{x\to a} dx = c\lim_{x\to a} x + d\lim_{x\to a} x$

Step 4: $c\lim_{x\to a} x + d\lim_{x\to a} x = ca + da$

$\lim_{x\to a}(cx+dx) = ca + da$

Finding the Limit of a Function

Given below are some of the steps involved in evaluating the limit of a function:

Step 1: Plug in the value of x in the given function.

Step 2: If the limit is definite or undefined, we are done with the limit of the function.

Step 3: If the limit is 0/0 form, we need to proceed further to evaluate the limit.

Step 4: Factorise the numerator and the denominator and simplify by dividing the numerator and the denominator by the common facto [i.e by removing the holes of the rational function].

Step 5: Plug in the value of x to find the limit of the function.

Step 6: If the function is of the form of special limits apply the value of the limits of the special limits and evaluate the limit of the given function.

Working Rule for Finding the Left and Right Hand Limits

To find the left hand limit as $x \to a$

1. Replace x as (a - h) in f(x)

2. Evaluate, $\lim_h \to 0(a-h)$

i.e $\lim_{x \to a^+} f(x) = \lim_{h \to 0} f(a+h)$

To find the right hand limit

1. Replace x as (x + h) in f(x)
2. Evaluate, $\lim_{h \to 0} f(a+h)$

$\lim_{x \to a^-} f(x) = \lim_{h \to 0} f(a+h)$

Examples on Finding the Limit of a Function

Let us discuss some examples on finding the limit of a function with different types.

Question 1: Find the limit of a function as x tends to 1

Solution:

We have the function $f(x) = \dfrac{x^4 - 1}{x - 1}$

When we plug in the value x = 1, we get $\dfrac{0}{0}$ form

Applying the special limit property (1), we get

$$\lim_{x \to a} \frac{x^4 - 1}{x - 1} = 4(1)^{4-1} = 4(1) = 4$$

Limit of the function is 4.

Question 2: Evaluate, $\lim_{x \to 2} \dfrac{x^2 - 5x + 6}{x^2 - 4}$

Solution:

We have $\lim_{x \to 2} \dfrac{x \ - 5x + 6}{x \ - 4}$ $\lim_{x \to 2} \dfrac{(x-2)(x-3)}{(x+2(x-2))}$

$\lim_{x \to 2} \dfrac{(x-3)}{(x+2)}$ [Dividing the numerator and the denominator by the common factor (x - 2)]

$= \dfrac{2-3}{2+2}$

$= -\dfrac{1}{4}$

limit of the function is $-\dfrac{1}{4}$

Question 3: Evaluate $\lim_{\theta\to 0}\frac{\sin\theta}{\theta}$

Solution:

We have

$\lim_{\theta\to 0}\frac{\sin\theta}{\theta} = \lim_{\theta\to 0}\frac{\sin\theta}{\theta}$ [multiplying the numerator and denominator by a]

$\lim_{\theta\to 0}\frac{\sin\theta}{a\theta}.a$

=(1)a

= a [using Special limit (2)]

Limit of the function is a.

Question 4: Find the limit of a function,

$$f(x)=\begin{cases}2x+3, ifx\le 2\\ 2x-3, ifx>2\end{cases}$$

Solution:

The left hand limit is $\lim_{x\to 2}-(2x+3)=2(2)+3=4+3=7$

The right hand limit is, $\lim_{x\to 2}+(2x+3)=2(2)-3=4-3=1$

Since the right hand limit is not equal to the left hand limit, the limit of the function does not exists.

Question 5: $\lim_{x\to 0}\frac{|\sin x|}{x}$

Solution:

we have, $\lim_{x\to 0}\frac{|\sin x|}{x}$

Since we have the modulus function, we us evaluate the left hand and right hand limits.

Left Hand limit = $\lim_{h\to 0}\frac{|\sin(0-h)|}{0-h}$

$$=\lim_{h\to 0}\frac{|\sin(-h)|}{-h}$$

$$= \lim_{h\to 0} \frac{\sin -h}{-h}$$

=-1

Right Hand limit $= \lim_{h\to 0} \frac{|\sin(0+h)|}{0+h}$

$$= \lim_{h\to 0} \frac{\sin(h)}{h} = 1$$

We observe that the left hand limit is not equal to the right hand limits.

Hence the limit does not exist.

Question 6: Evaluate $\lim_{x\to 0} \frac{x^3 \cot x}{1-\cos x}$

Solution:

We have $\lim_{x\to 0} \frac{x^3 \cot x}{1-\cos x}$

$$\lim_{x\to 0} \frac{x^3 \cos x}{(1-\cos x)} \frac{(1+\cos x)}{(1+\cos x)}$$

$$\lim_{x\to 0} \frac{x^3 \cos x(1+\cos x)}{\sin x(1-\cos^2 x)}$$

$$\lim_{x\to 0} \frac{x^3 \cos x(1+\cos x)}{\sin x(\sin^2 x)}$$

$$\lim_{x\to 0} \frac{x^3 \cos x(1+\cos x)}{(\sin^3 x)}$$

$$\frac{\lim_{x\to 0} \cos x(1+\cos x)}{\lim_{x\to 0} \frac{\sin^3 x}{x^3}}$$

$$= \frac{1+1}{1} = 2$$

Functions of a Single Variable

Suppose f : R → R is defined on the real line and p,L ∈ R. It is said the limit of f, as x approaches p, is L and written

$$\lim_{x\to p} f(x) = L,$$

if the following property holds:

- For every real $\epsilon > 0$, there exists a real $\delta > 0$ such that for all real x, $0 < |x - p| < \delta$ implies $|f(x) - L| < \epsilon$.

The value of the limit does not depend on the value of $f(p)$, nor even that p be in the domain of f.

A more general definition applies for functions defined on subsets of the real line. Let (a, b) be an open interval in R, and p a point of (a, b). Let f be a real-valued function defined on all of (a, b) except possibly at p itself. It is then said that the limit of f, as x approaches p, is L if, for every real $\epsilon > 0$, there exists a real $\delta > 0$ such that $0 < |x - p| < \delta$ and $x \in (a, b)$ implies $|f(x) - L| < \epsilon$.

Here again the limit does not depend on $f(p)$ being well-defined.

The letters ϵ and δ can be understood as "error" and "distance", and in fact Cauchy used ϵ as an abbreviation for "error" in some of his work, though in his definition of continuity he used an infinitesimal α rather than either ϵ or δ. In these terms, the error (ϵ) in the measurement of the value at the limit can be made as small as desired by reducing the distance (δ) to the limit point. As discussed below this definition also works for functions in a more general context. The idea that δ and ϵ represent distances helps suggest these generalizations.

Existence and One-sided Limits

Alternatively x may approach p from above (right) or below (left), in which case the limits may be written as

$$\lim_{x \to p^+} f(x) = L$$

or

$$\lim_{x \to p^-} f(x) = L$$

respectively. If these limits exist at p and are equal there, then this can be referred to as the limit of f(x) at p. If the one-sided limits exist at p, but are unequal, there is no limit at p (the limit at p does not exist). If either one-sided limit does not exist at p, the limit at p does not exist.

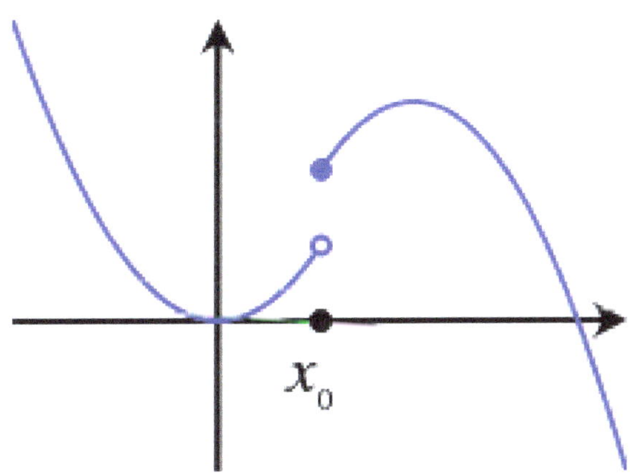

The limit as: $x \to x_0^+ \neq x \to x_0^-$. Therefore, the limit as $x \to x_0$ does not exist.

A formal definition is as follows. The limit of $f(x)$ as x approaches p from above is L if, for every $\epsilon > 0$, there exists a $\delta > 0$ such that $|f(x) - L| < \epsilon$ whenever $0 < x - p < \delta$. The limit of $f(x)$ as x approaches p from below is L if, for every $\epsilon > 0$, there exists a $\delta > 0$ such that $|f(x) - L| < \epsilon$ whenever $0 < p - x < \delta$.

If the limit does not exist then the oscillation of f at p is non-zero.

More General Subsets

Apart from open intervals, limits can be defined for functions on arbitrary subsets of R, as follows. Let f be a real-valued function defined on a subset S of the real line. Let p be a limit point of S—that is, p is the limit of some sequence of elements of S distinct from p. The limit of f, as x approaches p from values in S, is L if, for every $\epsilon > 0$, there exists a $\delta > 0$ such that $0 < |x - p| < \delta$ and $x \in S$ implies $|f(x) - L| < \epsilon$.

This limit is often written

$$L = \lim_{\substack{x \to p \\ x \in S}} f(x).$$

The condition that f be defined on S is that S be a subset of the domain of f. This generalization includes as special cases limits on an interval, as well as left-handed limits of real-valued functions (e.g., by taking S to be an open interval of the form $(-\infty, a)$), and right-handed limits (e.g., by taking S to be an open interval of the form (a, ∞)). It also extends the notion of one-sided limits to the included endpoints of (half-)closed intervals, so the Square root function $f(x) = \sqrt{x}$ can have limit 0 as x approaches 0 from above.

Deleted Versus Non-deleted Limits

The definition of limit given here does not depend on how (or whether) f is defined at p. Bartle (1967), refers to this as a *deleted limit*, because it excludes the value of f at p. The corresponding *non-deleted limit* does depend on the value of f at p, if p is in the domain of f:

- A number L is the non-deleted limit of f as x approaches p if, for every $\epsilon > 0$, there exists a $\delta > 0$ such that $| x - p | < \delta$ and $x \in Dm(f)$ implies $| f(x) - L | < \epsilon$.

The definition is the same, except that the neighborhood $| x - p | < \delta$ now includes the point p, in contrast to the (deleted) neighborhood $0 < | x - p | < \delta$. This makes the definition of a non-deleted limit less general. One of the advantages of working with non-deleted limits is that they allow to state the theorem about limits of compositions without any constraints on the functions (other than the existence of their non-deleted limits) (Hubbard (2015)).

Bartle (1967) notes that although by "limit" some authors do mean this non-deleted limit, deleted limits are the most popular. For example, Apostol (1974), Courant (1924), Hardy (1921), Rudin (1964), Whittaker & Watson (1902) all by "limit" mean the deleted version.

Examples

Non-existence of One-sided limit(s)

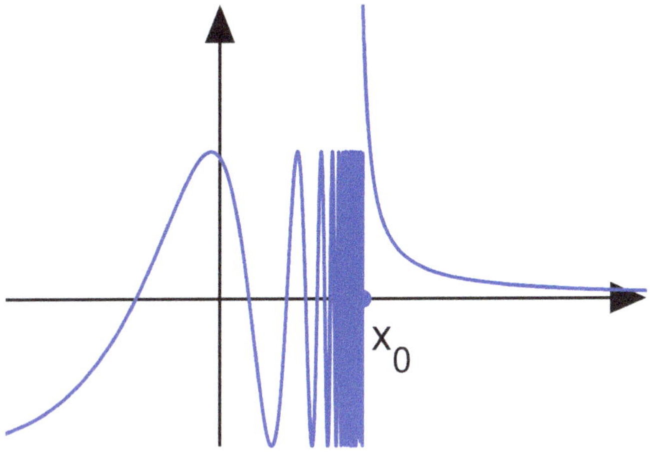

The function without a limit, at an essential discontinuity

The function

$$f(x)=\begin{cases}\sin\dfrac{5}{x-1} & \textit{for } x<1\\ 0 & \textit{for } x=1\\ \dfrac{0.1}{x-1} & \textit{for } x>1\end{cases}$$

has no limit at $x_0 = 1$ (the left-hand limit does not exist due to the oscillatory nature of the sine function, and the right-hand limit does not exist due to the asymptotic behaviour of the reciprocal function), but has a limit at every other *x*-coordinate.

The function

$$f(x)=\begin{cases}1 & x \textit{ rational}\\ 0 & x \textit{ irrational}\end{cases}$$

(the Dirichlet function) has no limit at any *x*-coordinate.

Non-equality of One-sided Limits

The function

$$f(x)=\begin{cases}1 & \textit{for } x<0\\ 2 & \textit{for } x\geq 0\end{cases}$$

has a limit at every non-zero *x*-coordinate (the limit equals 1 for negative *x* and equals 2 for positive *x*). The limit at $x = 0$ does not exist (the left-hand limit equals 1, whereas the right-hand limit equals 2).

Limits at only One Point

The functions

$$f(x) = \begin{cases} x & x \text{ rational} \\ 0 & x \text{ irrational} \end{cases}$$

and

$$f(x) = \begin{cases} |x| & x \text{ rational} \\ 0 & x \text{ irrational} \end{cases}$$

both have a limit at x = 0 and it equals 0.

Limits at Countably Many Points

The function

$$f(x) = \begin{cases} \sin x & x \text{ irrational} \\ 1 & x \text{ rational} \end{cases}$$

has a limit at any x-coordinate of the form $\frac{\pi}{2} + 2n\pi$, , where n is any integer.

Limit of a Function of Two Variables

If we have a function f(x,y) which is depends on two variables x and y. Then this given function has the limit say C as (x,y) → (a,b) provided that ϵ>0,∃δ>0 such that

$$|f(x,y) - C| < \epsilon \text{ whenever } 0 < \sqrt{(x-a)^2 + (y-b)^2} < \delta.$$

It is defined as $\lim_{(x,y)\to(a,b)} f(x,y) = C$

Limit of a Function of Two Variables Examples

Given below are some of the examples on limit of a function of two variables

Question 1: Evaluate $\lim_{x\to 2} \frac{x^3 - 8}{x^2 - 4}$

Solution:

$$\lim_{x\to 2} \frac{x^3 - 8}{x^2 - 4} = \lim_{x\to 2} \frac{x^3 - 2^3}{x^2 - 2^2}$$

$$= \lim_{x\to 2} \frac{x^3 - 2^3}{x - 2} \div \lim_{x\to 2} \frac{x^2 - 2^2}{x - 2}$$

$= 3(2)(3-1) \div 2(2)(2-1)$

$= 12 \div 4$

$= 3$

Question 2: Evaluate $\lim_{x \to 0} \sin(2x)$

Solution:

$$\lim_{x \to 0} \sin(2x) = \lim_{x \to 0} \left[\frac{\sin(2x)}{2} \times (2) \right]$$

$$= 2 \lim_{x \to 0} \frac{\sin(2x)}{2}$$

$$= 2 \times 1$$

$$= 2$$

Question 3: Evaluate $\lim_{x \to 0} \frac{x + 3\cos x}{2x - 3}$

Solution:

$$\lim_{x \to 0} \frac{x + 3\cos x}{2x - 3} = \frac{0 + 3(1)}{2(0) - 3} \text{ [as cos 0 = 1]}$$

$$= \frac{3}{-3}$$

$$= -1$$

Question 4: Simplify $\lim_{x \to 1} \left[x^2 - 4x + 2 \right]$

Solution:

$$\lim_{x \to 1} \left[x^2 - 4x + 2 \right]$$

Plugging $x = 1$, we get $f(1) = 1^2 - 4(1) + 2 = -1$

So, f(x) approaches -1 as x approaches 1.

Question 5: Simplify $\lim_{x \to 0} \frac{8x^3 - 27}{4x^2 + 6x + 9}$

Solution:

$$\lim_{x \to 0} \frac{8x^3 - 27}{4x^2 + 6x + 9} = \lim_{x \to 0} \frac{(2x)^3 - (3)^3}{4x^2 + 6x + 9}$$

Now, we use the formula, $a^3 - b^3 = (a-b)(a^2 + ab + b^2)$

$$\text{so}(2x)^3 - 3^3 = (2x-3)(4x^2+6x+9)$$

$$= \lim_{x\to 0} \frac{(2x-3)(4x^2+6x+9)}{4x^2+6x+9}$$

$$= \lim_{x\to 0}(2x-3)$$

Plugging x as 0, we get, $2(0)-3=-3$

So, f(x) approaches -3 as x approaches 0.

Functions on Metric Spaces

Suppose M and N are subsets of metric spaces A and B, respectively, and $f : M \to N$ is defined between M and N, with $x \in M$, p a limit point of M and $L \in N$. It is said that the limit of f as x approaches p is L and write

$$\lim_{x\to p} f(x) = L$$

if the following property holds:

- For every $\epsilon > 0$, there exists a $\delta > 0$ such that $d_B(f(x), L) < \epsilon$ whenever $0 < d_A(x, p) < \delta$.

Again, note that p need not be in the domain of f, nor does L need to be in the range of f, and even if $f(p)$ is defined it need not be equal to L.

An alternative definition using the concept of neighbourhood is as follows:

$$\lim_{x\to p} f(x) = L$$

if, for every neighbourhood V of L in B, there exists a neighbourhood U of p in A such that $f(U \cap M - \{p\}) \subseteq V$.

Functions on Topological Spaces

Suppose X, Y are topological spaces with Y a Hausdorff space. Let p be a limit point of $\Omega \subseteq X$, and $L \in Y$. For a function $f : \Omega \to Y$, it is said that the limit of f as x approaches p is L (i.e., $f(x) \to L$ as $x \to p$) and written

$$\lim_{x\to p} f(x) = L$$

if the following property holds:

- For every open neighborhood V of L, there exists an open neighborhood U of p such that $f(U \cap \Omega - \{p\}) \subseteq V$.

This last part of the definition can also be phrased "there exists an open punctured neighbourhood U of p such that $f(U \cap \Omega) \subseteq V$".

The domain of f does not need to contain p. If it does, then the value of f at p is irrelevant to the definition of the limit. In particular, if the domain of f is $X - \{p\}$ (or all of X), then the limit of f as $x \to p$ exists and is equal to L if, for all subsets Ω of X with limit point p, the limit of the restriction of f to Ω exists and is equal to L. Sometimes this criterion is used to establish the *non-existence* of the two-sided limit of a function on R by showing that the one-sided limits either fail to exist or do not agree. Such a view is fundamental in the field of general topology, where limits and continuity at a point are defined in terms of special families of subsets, called filters, or generalized sequences known as nets.

Alternatively, the requirement that Y be a Hausdorff space can be relaxed to the assumption that Y be a general topological space, but then the limit of a function may not be unique. In particular, one can no longer talk about *the limit* of a function at a point, but rather *a limit* or *the set of limits* at a point.

A function is continuous in a limit point p of and in its domain if and only if $f(p)$ is *the* (or, in the general case, *a*) limit of $f(x)$ as x tends to p.

Limits Involving Infinity

Limits at Infinity

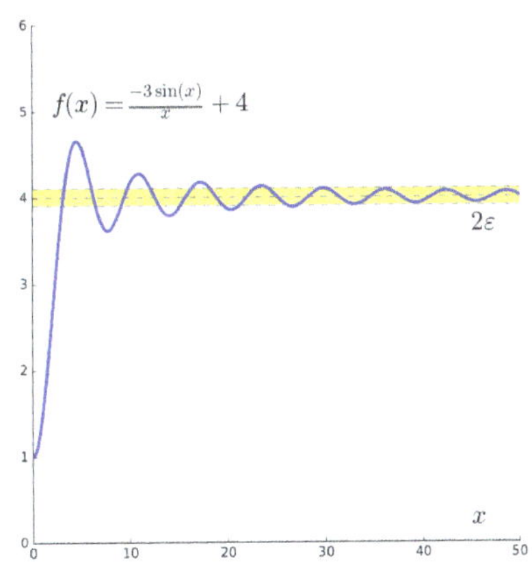

The limit of this function at infinity exists.

For $f(x)$ a real function, the limit of f as x approaches infinity is L, denoted

$$\lim_{x \to \infty} f(x) = L,$$

means that for all $\varepsilon > 0$, there exists c such that $|f(x) - L| < \varepsilon$ whenever $x > c$. Or, symbolically:

$$\forall \varepsilon > 0 \; \exists c \; \forall x > c : |f(x) - L| < \varepsilon.$$

Similarly, the limit of f as x approaches negative infinity is L, denoted

$$\lim_{x \to -\infty} f(x) = L,$$

means that for all $\varepsilon > 0$ there exists c such that $|f(x) - L| < \varepsilon$ whenever $x < c$. Or, symbolically:

$$\forall \varepsilon > 0 \, \exists c \, \forall x < c : | f(x) - L | < \varepsilon .$$

For example

$$\lim_{x \to -\infty} e^x = 0 .$$

Infinite Limits

Limits can also have infinite values. When infinities are not considered legitimate values, which is standard, a formalist will insist upon various circumlocutions. For example, rather than say that a limit *is* infinity, the *proper* thing is to say that the function "diverges" or "grows without bound". In particular, the following informal example of how to pronounce the notation is arguably inappropriate in the classroom (or any other formal setting). In any case, for example the limit of f as x approaches a is infinity, denoted

$$\lim_{x \to a} f(x) = \infty,$$

means that for all $\varepsilon > 0$ there exists $\delta > 0$ such that $f(x) > \varepsilon$ whenever $|x - a| < \delta$.

These ideas can be combined in a natural way to produce definitions for different combinations, such as

$$\lim_{x \to \infty} f(x) = \infty, \lim_{x \to a^+} f(x) = -\infty.$$

For example

$$\lim_{x \to 0^+} \ln x = -\infty.$$

Limits involving infinity are connected with the concept of asymptotes.

These notions of a limit attempt to provide a metric space interpretation to limits at infinity. However, note that these notions of a limit are consistent with the topological space definition of limit if

- a neighborhood of $-\infty$ is defined to contain an interval $[-\infty, c)$ for some $c \in R$
- a neighborhood of ∞ is defined to contain an interval $(c, \infty]$ where $c \in R$
- a neighborhood of $a \in R$ is defined in the normal way metric space R

In this case, $\overline{\mathrm{R}}$ is a topological space and any function of the form $f: X \to Y$ with $X, Y \subseteq \overline{\mathrm{R}}$ is subject to the topological definition of a limit. Note that with this topological definition, it is easy to define infinite limits at finite points, which have not been defined above in the metric sense.

Alternative Notation

Many authors allow for the projectively extended real line to be used as a way to include infinite values as well as extended real line. With this notation, the extended real line is given as $\mathrm{R} \cup \{-\infty,$

$+\infty\}$ and the projectively extended real line is $\mathbb{R} \cup \{\infty\}$ where a neighborhood of ∞ is a set of the form $\{x: |x| > c\}$. The advantage is that one only needs three definitions for limits (left, right, and central) to cover all the cases. As presented above, for a completely rigorous account, we would need to consider 15 separate cases for each combination of infinities (five directions: $-\infty$, left, central, right, and $+\infty$; three bounds: $-\infty$, finite, or $+\infty$). There are also noteworthy pitfalls. For example, when working with the extended real line, x^{-1} does not possess a central limit (which is normal):

$$\lim_{x\to 0^+} \frac{1}{x} = +\infty, \lim_{x\to 0^-} \frac{1}{x} = -\infty.$$

In contrast, when working with the projective real line, infinities (much like 0) are unsigned, so, the central limit *does* exist in that context:

$$\lim_{x\to 0^+} \frac{1}{x} = \lim_{x\to 0^-} \frac{1}{x} = \lim_{x\to 0} \frac{1}{x} = \infty.$$

In fact there are a plethora of conflicting formal systems in use. In certain applications of numerical differentiation and integration, it is, for example, convenient to have signed zeroes. A simple reason has to do with the converse of $\lim_{x\to 0^-} x^{-1} = -\infty$, namely, it is convenient for $\lim_{x\to -\infty} x^{-1} = -0$ to be considered true. Such zeroes can be seen as an approximation to infinitesimals.

Limits at Infinity for Rational Functions

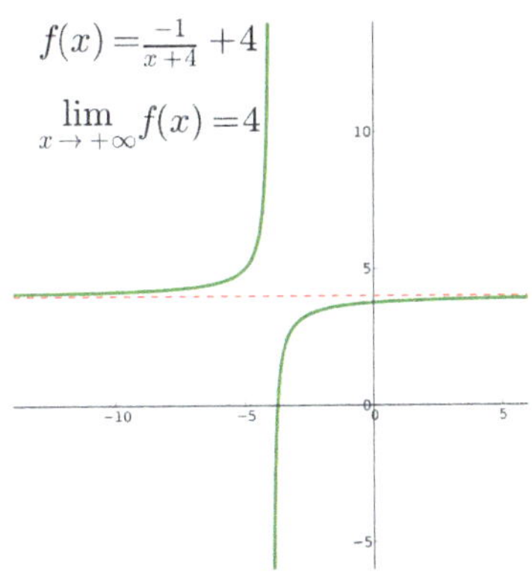

Horizontal asymptote about $y = 4$

There are three basic rules for evaluating limits at infinity for a rational function $f(x) = p(x)/q(x)$: (where p and q are polynomials):

- If the degree of p is greater than the degree of q, then the limit is positive or negative infinity depending on the signs of the leading coefficients;
- If the degree of p and q are equal, the limit is the leading coefficient of p divided by the leading coefficient of q;
- If the degree of p is less than the degree of q, the limit is 0.

If the limit at infinity exists, it represents a horizontal asymptote at $y = L$. Polynomials do not have horizontal asymptotes; such asymptotes may however occur with rational functions.

Functions of more than One Variable

By noting that $|x - p|$ represents a distance, the definition of a limit can be extended to functions of more than one variable. In the case of a function $f: R^2 \to R$,

$$\lim_{(x,y)\to(p,q)} f(x,y) = L$$

if

> for every $\epsilon > 0$ there exists a $\delta > 0$ such that for all (x,y) with $0 < ||(x,y) - (p,q)|| < \delta$, then $|f(x,y) - L| < \epsilon$ where $||(x,y) - (p,q)||$ represents the Euclidean distance. This can be extended to any number of variables.

Sequential Limits

Let $f: X \to Y$ be a mapping from a topological space X into a Hausdorff space Y, $p \in X$ a limit point of X and $L \in Y$.

> The sequential limit of f as x tends to p is L if, for every sequence (x_n) in $X - \{p\}$ that converges to p, the sequence $f(x_n)$ converges to L.

If L is the limit (in the sense above) of f as x approaches p, then it is a sequential limit as well, however the converse need not hold in general. If in addition X is metrizable, then L is the sequential limit of f as x approaches p if and only if it is the limit (in the sense above) of f as x approaches p.

Other Characterizations

In Terms of Sequences

For functions on the real line, one way to define the limit of a function is in terms of the limit of sequences. (This definition is usually attributed to Eduard Heine.) In this setting:

$$\lim_{x\to a} f(x) = L$$

if and only if for all sequences x_n (with x_n not equal to a for all n) converging to a the sequence $f(x_n)$ converges to L. It was shown by Sierpiński in 1916 that proving the equivalence of this definition and the definition above, requires and is equivalent to a weak form of the axiom of choice. Note that defining what it means for a sequence x_n to converge to a requires the epsilon, delta method.

Similarly as it was the case of Weierstrass's definition, a more general Heine definition applies to functions defined on subsets of the real line. Let f be a real-valued function with the domain $Dm(f)$. Let a be the limit of a sequence of elements of $Dm(f) \setminus \{a\}$. Then the limit (in this sense) of f is L as x approaches p if for every sequence $x_n \in Dm(f) \setminus \{a\}$ (so that for all n, x_n is not equal to a) that converges to a, the sequence $f(x_n)$ converges to L. This is the same as the definition of a sequential limit in the preceding section obtained by regarding the subset $Dm(f)$ of R as a metric space with the induced metric.

In Non-standard Calculus

In non-standard calculus the limit of a function is defined by:

$$\lim_{x \to a} f(x) = L$$

if and only if for all $x \in \mathbb{R}^*$, $f^*(x) - L$ is infinitesimal whenever $x - a$ is infinitesimal. Here $\mathbb{R}^*$ are the hyperreal numbers and f^* is the natural extension of *f* to the non-standard real numbers. Keisler proved that such a hyperreal definition of limit reduces the quantifier complexity by two quantifiers. On the other hand, Hrbacek writes that for the definitions to be valid for all hyperreal numbers they must implicitly be grounded in the ϵ-δ method, and claims that, from the pedagogical point of view, the hope that non-standard calculus could be done without ϵ-δ methods cannot be realized in full. Błaszczyk et al. detail the usefulness of microcontinuity in developing a transparent definition of uniform continuity, and characterize Hrbacek's criticism as a "dubious lament".

In Terms of Nearness

At the 1908 international congress of mathematics F. Riesz introduced an alternate way defining limits and continuity in concept called "nearness". A point x is defined to be near a set $A \subseteq \mathbb{R}$ if for every $r > 0$ there is a point $a \in A$ so that $|x - a| < r$. In this setting the

$$\lim_{x \to a} f(x) = L$$

if and only if for all $A \subseteq \mathbb{R}$, L is near $f(A)$ whenever a is near A. Here $f(A)$ is the set $\{f(x) \mid x \in A\}$. This definition can also be extended to metric and topological spaces.

Relationship to Continuity

The notion of the limit of a function is very closely related to the concept of continuity. A function f is said to be continuous at c if it is both defined at c and its value at c equals the limit of *f* as x approaches c:

$$\lim_{x \to c} f(x) = f(c).$$

(We have here assumed that c is a limit point of the domain of *f*.)

Properties

If a function *f* is real-valued, then the limit of *f* at *p* is *L* if and only if both the right-handed limit and left-handed limit of *f* at *p* exist and are equal to *L*.

The function *f* is continuous at *p* if and only if the limit of *f*(*x*) as *x* approaches *p* exists and is equal to *f*(*p*). If $f : M \to N$ is a function between metric spaces *M* and *N*, then it is equivalent that *f* transforms every sequence in *M* which converges towards *p* into a sequence in *N* which converges towards *f*(*p*).

If N is a normed vector space, then the limit operation is linear in the following sense: if the limit of $f(x)$ as x approaches p is L and the limit of $g(x)$ as x approaches p is P, then the limit of $f(x) + g(x)$ as x approaches p is $L + P$. If a is a scalar from the base field, then the limit of $af(x)$ as x approaches p is aL.

If f is a real-valued (or complex-valued) function, then taking the limit is compatible with the algebraic operations, *provided* the limits on the *right* sides of the equations below exist (the last identity only holds if the denominator is non-zero). This fact is often called the algebraic limit theorem.

$$\begin{aligned}
\lim_{x\to p} (f(x)+g(x)) &= \lim_{x\to p} f(x) + \lim_{x\to p} g(x) \\
\lim_{x\to p} (f(x)-g(x)) &= \lim_{x\to p} f(x) - \lim_{x\to p} g(x) \\
\lim_{x\to p} (f(x)\cdot g(x)) &= \lim_{x\to p} f(x) \cdot \lim_{x\to p} g(x) \\
\lim_{x\to p} (f(x)/g(x)) &= \lim_{x\to p} f(x) / \lim_{x\to p} g(x)
\end{aligned}$$

In each case above, when the limits on the right do not exist, or, in the last case, when the limits in both the numerator and the denominator are zero, nonetheless the limit on the left, called an *indeterminate form*, may still exist—this depends on the functions f and g. These rules are also valid for one-sided limits, for the case $p = \pm\infty$, and also for infinite limits using the rules

- $q + \infty = \infty$ for $q \neq -\infty$
- $q \times \infty = \infty$ if $q > 0$
- $q \times \infty = -\infty$ if $q < 0$
- $q / \infty = 0$ if $q \neq \pm \infty$

Note that there is *no* general rule for the case $q / 0$; it all depends on the way 0 is approached. Indeterminate forms—for instance, $0/0$, $0\times\infty$, $\infty-\infty$, and ∞/∞—are also not covered by these rules, but the corresponding limits can often be determined with L'Hôpital's rule or the Squeeze theorem.

Limits of Compositions of Functions

In general, from knowing that

$$\lim_{y\to b} f(y) = c \text{ and } \lim_{x\to a} g(x) = b,$$

it does not follow that $\lim_{x\to a} f(g(x)) = c$. However, this "chain rule" does hold if one of the following additional conditions holds:

- $f(b) = c$ (that is, f is continuous at b), or
- g does not take the value b near a (that is, there exists a $\delta > 0$ such that if $0 < |x-a| < \delta$ then $|g(x)-b| > 0$).

As an example of this phenomenon, consider the following functions that violates both additional restrictions:

$$f(x) = g(x) = \begin{cases} 0 & if\, x \neq 0 \\ 1 & if\, x = 0 \end{cases}.$$

Since the value at $f(0)$ is a removable discontinuity,

$$\lim_{x \to a} f(x) = 0 \text{ for all } a.$$

Thus, the naïve chain rule would suggest that the limit of $f(f(x))$ is 0. However, it is the case that

$$f(f(x)) = \begin{cases} 1 & if\, x \neq 0 \\ 0 & if\, x = 0 \end{cases}$$

and so

$$\lim_{x \to a} f(f(x)) = 1 \text{ for all } a.$$

Limits of Special Interest

Rational Functions

For n a nonnegative integer and constants $a_1, a_2, a_3, \ldots, a_n$ and $b_1, b_2, b_3, \ldots, b_n$,

- $$\lim_{x \to \infty} \frac{a_1 x^n + a_2 x^{n-1} + a_3 x^{n-2} + \ldots + a_n}{b_1 x^n + b_2 x^{n-1} + b_3 x^{n-2} + \ldots + b_n} = \frac{a_1}{b_1}$$

This can be proven by dividing both the numerator and denominator by x^n. If the numerator is a polynomial of higher degree, the limit does not exist. If the denominator is of higher degree, the limit is 0.

Trigonometric Functions

- $$\lim_{x \to 0} \frac{\sin x}{x} = 1$$
- $$\lim_{x \to 0} \frac{1 - \cos x}{x} = 0$$
- $$\lim_{x \to \infty} x \sin\left(\frac{1}{x}\right) = 1$$

Exponential Functions

- $$\lim_{x \to 0} (1 + x)^{\frac{1}{x}} = \lim_{r \to \infty} \left(1 + \frac{1}{r}\right)^r = e$$

- $\lim_{x\to 0}\frac{e^x - 1}{x} = 1$
- $\lim_{x\to 0}\frac{e^{ax} - 1}{bx} = \frac{a}{b}$
- $\lim_{x\to 0}\frac{c^{ax} - 1}{bx} = \frac{a}{b}\ln c$
- $\lim_{x\to 0^+} x^x = 1$

Logarithmic Functions

- $\lim_{x\to 0}\frac{\ln(1+x)}{x} = 1$
- $\lim_{x\to 0}\frac{\ln(1+ax)}{bx} = \frac{a}{b}$
- $\lim_{x\to 0}\frac{\log_c(1+ax)}{bx} = \frac{a}{b\ln c}$

L'Hôpital's Rule

This rule uses derivatives to find limits of indeterminate forms 0/0 or ±∞/∞, and only applies to such cases. Other indeterminate forms may be manipulated into this form. Given two functions $f(x)$ and $g(x)$, defined over an open interval I containing the desired limit point c, then if:

1. $\lim_{x\to c} f(x) = \lim_{x\to c} g(x) = 0,$ or $\lim_{x\to c} f(x) = \pm\lim_{x\to c} g(x) = \pm\infty,$, and
2. f and g are differentiable over $I \setminus \{c\}$, and
3. $g'(x) \neq 0$ for all $x \in I \setminus \{c\}$, and
4. $\lim_{x\to c}\frac{f'(x)}{g'(x)}$ exists,

then:

$$\lim_{x\to c}\frac{f(x)}{g(x)} = \lim_{x\to c}\frac{f'(x)}{g'(x)}$$

Normally, the first condition is the most important one.

For example $\lim_{x\to 0}\frac{\sin(2x)}{\sin(3x)} = \lim_{x\to 0}\frac{2\cos(2x)}{3\cos(3x)} = \frac{2\cdot 1}{3\cdot 1} = \frac{2}{3}.$

Summations and Integrals

Specifying an infinite bound on a summation or integral is a common shorthand for specifying a limit.

A short way to write the limit $\lim_{n\to\infty}\sum_{i=s}^{n} f(i)$ is $\sum_{i=s}^{\infty} f(i)$. An important example of limits of sums such as these are series.

A short way to write the limit $\lim_{x\to\infty}\int_a^x f(t)\,dt$ is $\int_a^{\infty} f(t)\,dt$.

A short way to write the limit $\lim_{x\to-\infty}\int_x^b f(t)\,dt$ is $\int_{-\infty}^{b} f(t)\,dt$.

Continuous Function

In calculus, a continuous function is a real-valued function whose graph does not have any breaks or holes. Continuity lays the foundational groundwork for the intermediate value theorem and extreme value theorem. They are in some sense the "nicest" functions possible, and many proofs in real analysis rely on approximating arbitrary functions by continuous functions.

In calculus, knowing if the function is continuous is essential, because differentiation is only possible when the function is continuous. The concept of continuity is simple: If the graph of the function doesn't have any breaks or holes in it within a certain interval, the function is said to be continuous over that interval. Thus, simply drawing the graph might tell you if the function is continuous or not. However not all functions are easy to draw, and sometimes we will need to use the definition of continuity to determine a function's continuity. The mathematical definition of a continuous function is as follows:

For a function $f(x)$ to be continuous at a point $x = a$, it must satisfy all three of the following conditions:

(i) $f(a)$ exists.

(ii) $\lim_{x\to a} f(x)$ exists.

(iii) $\lim_{x\to a} f(x) = f(a)$.

When we say a function f is continuous on [a,b], it means that for all elements in the interval, the above conditions are satisfied. But do note that for while trying to prove continuity on [a,b], we don't have to take into account LHL for 'a' and RHL for 'b' as the points to the left of a and to the right of b are not included in [a,b]. We only consider RHL for 'a' and LHL for 'b'.

Examples

Here are a few example problems. Using the definition above, try to determine if they are continuous or not.

Is the function $f(x)=\begin{cases}2x+1 & (x<3)\\ 3x-2 & (x\geq 3)\end{cases}$ continuous for all $x\in\mathbb{R}$?

We know that the graphs of $y=2x+1$ and $y=3x-2$ are continuous, so we only need to see if the function is continuous at $x=3$. The procedure is simply using the definition above, as follows:

(i) Since $f(3)=3\times3-2=7, f(3)$ exists.

(ii) In order to see whether the limit exists or not, we have to check the limit from both sides. The left-hand and right-hand limits are

$$\lim_{x\to3^-}f(x)=\lim_{x\to3^-}(2x+1)=2\times3+1=7 \text{ and } \lim_{x\to3^+}f(x)=\lim_{x\to3^+}(3x-2)=3\times3-2=7$$

respectively. Because the limits from both sides are equal, $\lim_{x\to3}f(x)$ exists.

(iii) Now from (i) and (ii), we have $\lim_{x\to3}f(x)=f(3)=7,$ so the function is continuous at $x=3.$

Example

Is the function $f(x)=\begin{cases}x+1 & (x<2)\\ x^2 & (x=2)\\ 2x-1 & (x>2)\end{cases}$ continuous for all $x\in\mathbb{R}$?

We know that the graphs of $y=x+1$, $y=x^2$, and $y=2x-1$ are continuous, so we only need to see if the function is continuous at $x=2.$ Again we use the same procedure, as shown below:

(i) Since $f(2)=2^2=4, f(2)$ exists.

(ii) The left-hand and right-hand limits are

$$\lim_{x\to2^-}f(x)=\lim_{x\to2^-}(x+1)=3$$
$$\lim_{x\to2^+}f(x)=\lim_{x\to2^+}(2x-1)=3,$$

respectively. Therefore, $\lim_{x\to2^-}f(x)=\lim_{x\to2^+}f(x)=\lim_{x\to3}f(x)=3.$

(iii) Now from (i) and (ii), we have $\lim_{x\to2}f(x)\neq f(2),$ so the function is not continuous at $x=2.$

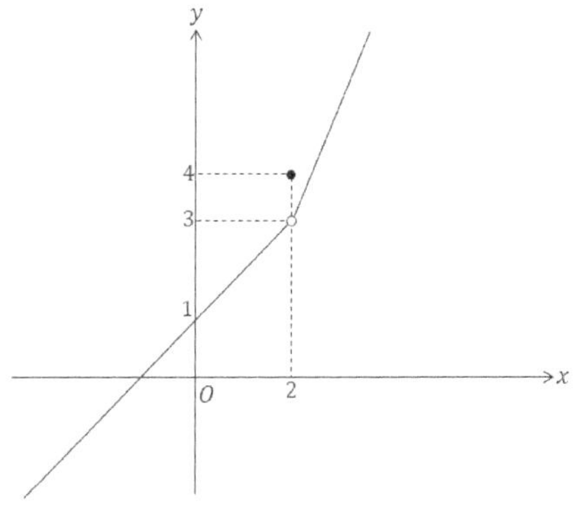

The graph of the function would look like the figure above. Observe that there is a "hole" at $x = 2$, which causes the discontinuity.

Example

Is the function $f(x) = \begin{cases} -x^3 + x + 1 (x \le 1) \\ 2x^2 + 3x - 2 (x > 1) \end{cases}$ continuous for all $x \in \mathbb{R}$?

We know that the graphs of $y = -x^3 + x + 1$ and $y = 2x^2 + 3x - 2$ are continuous, so we only need to see if the function is continuous at $x = 1$

(i) Since $f(1) = 1$, $f(1)$ exists.

(ii) The left-hand and right-hand limits are

$$\lim_{x \to 1^-} f(x) = \lim_{x \to 1^-} (-x^3 + x + 1) = 1$$
$$\lim_{x \to 1^+} f(x) = \lim_{x \to 1^+} (2x^2 + 3x - 2) = 3,$$

respectively. Since the left-hand limit and right-hand limit are not equal, $\lim_{x \to 1} f(x)$ does not exist, so the function $f(x)$ is not continuous at $x = 1$.

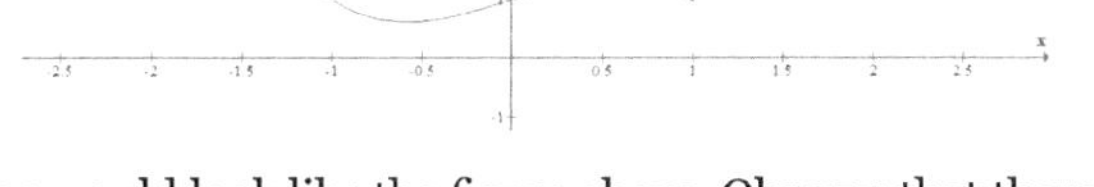

The graph of the function would look like the figure above. Observe that there is a "break" at $x = 1$, which causes the discontinuity.

Example

Is the function $f(x) = \begin{cases} -\cos x & (x < 0) \\ e^x - 2 & (x \ge 0) \end{cases}$ continuous for all $x \in \mathbb{R}$?

We know that the graphs of $y = -\cos x$ and $y = e^x - 2$ are continuous, so we only need to see if the function is continuous at $x = 0$.

(i) Since $f(0) = e^0 - 2 = -1$, $f(0)$ exists.

(ii) The left-hand and right-hand limits are

$$\lim_{x\to 0^-} f(x) = \lim_{x\to 0^-} (-\cos x) = -1$$

$$\lim_{x\to 0^+} f(x) = \lim_{x\to 0^+} (e^x - 2) = -1,$$

respectively. Therefore, $\lim_{x\to 0^-} f(x) = \lim_{x\to 0^+} f(x) = \lim_{x\to 0}(x) = -1.$

(iii) Now from (i) and (ii), we have $\lim_{x\to 2} f(x) = f(2) = -1$, so the function is continuous at $x = 0$.

Real Functions

A real function, that is a function from real numbers to real numbers can be represented by a graph in the Cartesian plane; such a function is continuous if, roughly speaking, the graph is a single unbroken curve whose domain is the entire real line. A more mathematically rigorous definition is given below.

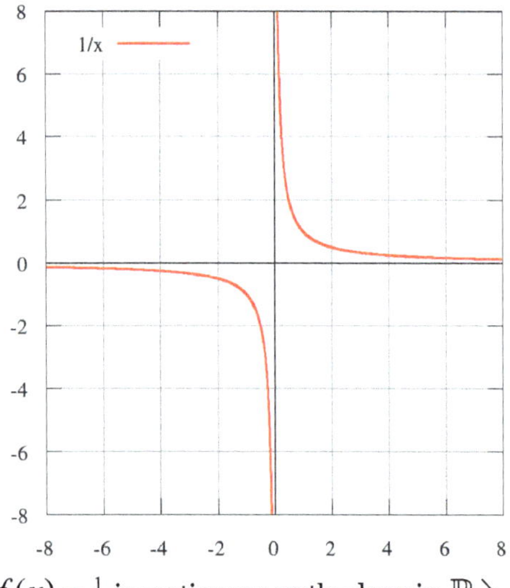

The function $f(x) = \frac{1}{x}$ is continuous on the domain $\mathbb{R} \setminus \{0\}$, but is not continuous over the domain $\mathbb{R}$ because it is undefined at $x = 0$

A rigorous definition of continuity of real functions is usually given in a first course in calculus in terms of the idea of a limit. First, a function f with variable x is said to be continuous *at the point* c on the real line, if the limit of $f(x)$, as x approaches that point c, is equal to the value $f(c)$; and second, the *function (as a whole)* is said to be *continuous*, if it is continuous at every point. A function is said to be *discontinuous* (or to have a *discontinuity*) at some point when it is not continuous there. These points themselves are also addressed as *discontinuities*.

There are several different definitions of continuity of a function. Sometimes a function is said to be continuous if it is continuous at every point in its domain. In this case, the function $f(x) = \tan(x)$, with the domain of all real $x \neq (2n+1)\pi/2$, n any integer, is continuous. Sometimes an exception is made for boundaries of the domain. For example, the graph of the function $f(x) = \sqrt{x}$, with the domain of all non-negative reals, has a *left-hand* endpoint. In this case only the limit from the *right* is required to equal the value of the function. Under this definition f is continuous at the boundary $x = 0$ and so for all non-negative arguments. The most common and restrictive definition is that a function is continuous if it is continuous at all real numbers. In this case, the previous two examples are not continuous, but every polynomial function is continuous, as are the sine, cosine, and exponential functions. Care should be exercised in using the word *continuous*, so that it is clear from the context which meaning of the word is intended.

Using mathematical notation, there are several ways to define continuous functions in each of the three senses mentioned above.

Let

$f : D \to \mathbf{R}$ be a function defined on a subset D of the set $\mathbf{R}$ of real numbers.

This subset D is the domain of f. Some possible choices include

$D = \mathbf{R}$ (D is the whole set of real numbers), or, for a and b real numbers,

$D = [a,b] = \{x \in \mathbf{R} | a \leq x \leq b\}$ (D is a closed interval), or

$D = (a,b) = \{x \in \mathbf{R} | a < x < b\}$ (D is an open interval).

In case of the domain D being defined as an open interval, a and b are no boundaries in the above sense and the values of $f(a)$ and $f(b)$ do not matter for continuity on D.

Definition in Terms of Limits of Functions

The function f is *continuous at some point* c of its domain if the limit of $f(x)$, as x approaches c through the domain of f, exists and is equal to $f(c)$. In mathematical notation, this is written as

$$\lim_{x \to c} f(x) = f(c).$$

In detail this means three conditions: first, f has to be defined at c. Second, the limit on the left hand side of that equation has to exist. Third, the value of this limit must equal $f(c)$.

(We have here assumed that the domain of f does not have any isolated points. For example, an interval or union of intervals has no isolated points.)

Definition in Terms of Neighborhoods

A neighborhood of a point c is a set that contains all points of the domain within some fixed distance of c. Intuitively, a function is continuous at a point c if the range of the restriction of f to a neighborhood of c shrinks to a single point $f(c)$ as the width of the neighborhood around c shrinks to zero. More precisely, a function *f* is continuous at a point *c* of its domain if, for any neighborhood $N_1(f(c))$ there is a neighborhood $N_2(c)$ such that $f(x) \in N_1(f(c))$ whenever $x \in N_2(c)$.

This definition only requires that the domain and the codomain are topological spaces and is thus the most general definition. It follows from this definition that a function *f* is automatically continuous at every isolated point of its domain. As a specific example, every real valued function on the set of integers is continuous.

Definition in Terms of Limits of Sequences

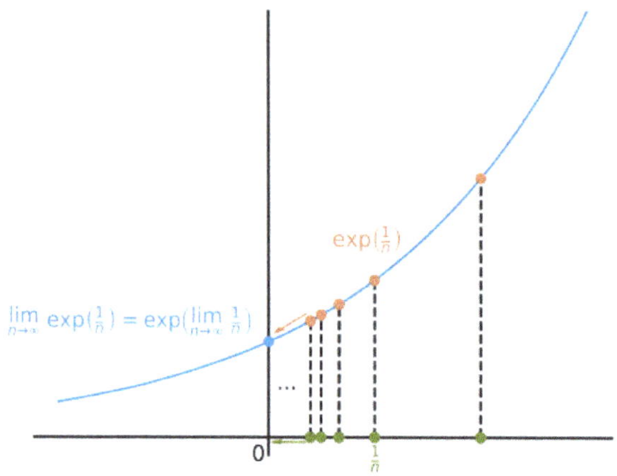

The sequence exp(1/*n*) converges to exp(0)

One can instead require that for any sequence $(x_n)_{n\in\mathbb{N}}$ of points in the domain which converges to *c*, the corresponding sequence $\left(f(x_n)\right)_{n\in\mathbb{N}}$ converges to *f*(*c*). In mathematical notation, $\forall (x_n)_{n\in\mathbb{N}} \subset D : \lim_{n\to\infty} x_n = c \Rightarrow \lim_{n\to\infty} f(x_n) = f(c).$

Weierstrass and Jordan Definitions (Epsilon–delta) of Continuous Functions

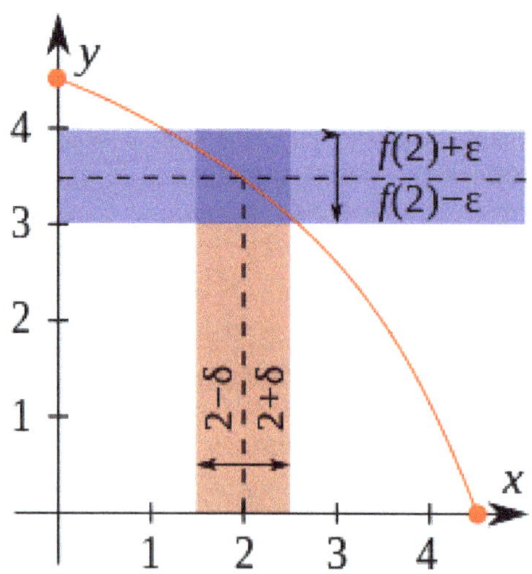

Illustration of the ϵ-δ-definition: for ϵ=0.5, c=2, the value δ=0.5 satisfies the condition of the definition.

Explicitly including the definition of the limit of a function, we obtain a self-contained definition: Given a function $f : D \rightarrow R$ as above and an element x_0 of the domain *D*, *f* is said to be continuous at the point x_0 when the following holds: For any number $\epsilon > 0$, however small, there exists some number $\delta > 0$ such that for all x in the domain of f with $x_0 - \delta < x < x_0 + \delta$, the value of $f(x)$ satisfies

$$f(x_0) - \varepsilon < f(x) < f(x_0) + \varepsilon.$$

Alternatively written, continuity of $f: D \to R$ at $x_0 \in D$ means that for every $\epsilon > 0$ there exists a $\delta > 0$ such that for all $x \in D$:

$$|x - x_0| < \delta \Rightarrow |f(x) - f(x_0)| < \varepsilon.$$

More intuitively, we can say that if we want to get all the $f(x)$ values to stay in some small neighborhood around $f(x_0)$, we simply need to choose a small enough neighborhood for the x values around x_0. If we can do that no matter how small the $f(x)$ neighborhood is, then f is continuous at x_0.

In modern terms, this is generalized by the definition of continuity of a function with respect to a basis for the topology, here the metric topology.

Weierstrass had required that the interval $x_0 - \delta < x < x_0 + \delta$ be entirely within the domain D, but Jordan removed that restriction.

Definition in Terms of Control of the Remainder

In proofs and numerical analysis we often need to know how fast limits are converging, or in other words, control of the remainder. We can formalise this to a definition of continuity. A function $C: [0, \infty) \to [0, \infty]$ is called a control function if

- C is non decreasing
- $\inf_{\delta > 0} C(\delta) = 0$

A function $f: D \to R$ is C-continuous at x_0 if

$$|f(x) - f(x_0)| \leq C(|x - x_0|) \text{ for all } x \in D$$

A function is continuous in x_0 if it is C-continuous for some control function C.

This approach leads naturally to refining the notion of continuity by restricting the set of admissible control functions. For a given set of control functions $\mathcal{C}$ a function is $\mathcal{C}$-continuous if it is $\mathcal{C}$-continuous for some $C \in \mathcal{C}$. For example the Lipschitz and Hölder continuous functions of exponent α below are defined by the set of control functions

$$\mathcal{C}_{Lipschitz} = \{C \mid C(\delta) = K|\delta|, K > 0\}$$

respectively

$$\mathcal{C}_{Hölder-\alpha} = \{C \mid C(\delta) = K|\delta|^{\alpha}, K > 0\}.$$

Definition using Oscillation

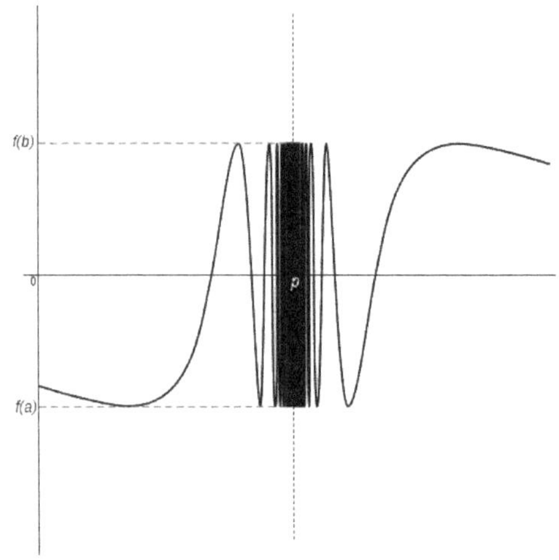

The failure of a function to be continuous at a point is quantified by its oscillation.

Continuity can also be defined in terms of oscillation: a function *f* is continuous at a point x_0 if and only if its oscillation at that point is zero; in symbols, $\omega_f(x_0) = 0$. A benefit of this definition is that it *quantifies* discontinuity: the oscillation gives how *much* the function is discontinuous at a point.

This definition is useful in descriptive set theory to study the set of discontinuities and continuous points – the continuous points are the intersection of the sets where the oscillation is less than ϵ (hence a G_δ set) – and gives a very quick proof of one direction of the Lebesgue integrability condition.

The oscillation is equivalent to the ϵ-δ definition by a simple re-arrangement, and by using a limit (lim sup, lim inf) to define oscillation: if (at a given point) for a given ϵ_0 there is no δ that satisfies the ϵ-δ definition, then the oscillation is at least ϵ_0, and conversely if for every ϵ there is a desired δ, the oscillation is 0. The oscillation definition can be naturally generalized to maps from a topological space to a metric space.

Definition using the Hyperreals

Cauchy defined continuity of a function in the following intuitive terms: an infinitesimal change in the independent variable corresponds to an infinitesimal change of the dependent variable. Non-standard analysis is a way of making this mathematically rigorous. The real line is augmented by the addition of infinite and infinitesimal numbers to form the hyperreal numbers. In nonstandard analysis, continuity can be defined as follows.

> A real-valued function *f* is continuous at *x* if its natural extension to the hyperreals has the property that for all infinitesimal *dx*, $f(x+dx) - f(x)$ is infinitesimal.

In other words, an infinitesimal increment of the independent variable always produces to an infinitesimal change of the dependent variable, giving a modern expression to Augustin-Louis Cauchy's definition of continuity.

Construction of Continuous Functions

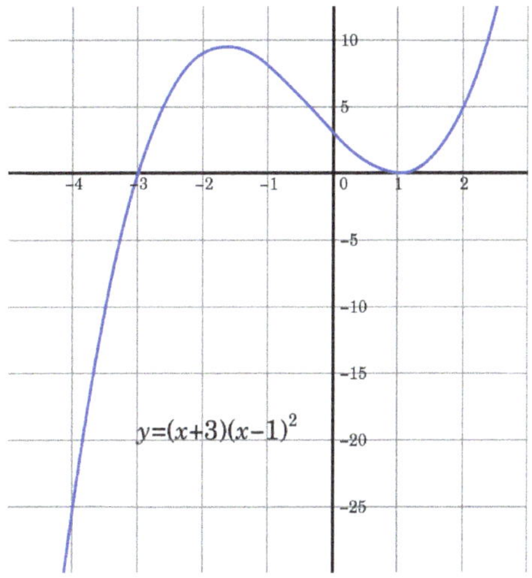

The graph of a cubic function has no jumps or holes. The function is continuous.

Checking the continuity of a given function can be simplified by checking one of the above defining properties for the building blocks of the given function. It is straightforward to show that the sum of two functions, continuous on some domain, is also continuous on this domain. Given

$$f, g : D \to \mathbf{R}$$

then the sum of continuous functions

$$s = f + g, \text{ defined by } s(x) = f(x) + g(x) \text{ for all } x \in D$$

is continuous in D.

The same holds for the product of continuous functions.

$$p = f \cdot g, \text{ defined by } p(x) = f(x) \cdot g(x) \text{ for all } x \in D,$$

is continuous in D.

Combining the above preservations of continuity and the continuity of constant functions and of the identity function $I(x) = x$ on $\mathbf{R}$, one arrives at the continuity of all polynomial functions on $\mathbf{R}$, such as

$$f(x) = x^3 + x^2 - 5x + 3$$

In the same way it can be shown that the reciprocal of a continuous function

$$r = 1/f, \text{ defined by } r(x) = 1/f(x) \text{ for all } x \in D \text{ such that } f(x) \neq 0,$$

is continuous in $D \setminus \{x : g(x) = 0\}$.

This implies that, excluding the roots of g, the quotient of continuous functions

$q = f / g$, defined by $g(x) = f(x)/g(x)$ for all $x \in D$, such that $g(x) \neq 0$,

is also continuous on $D \setminus \{x : g(x) = 0\}$.

For example, the function (pictured)

$$y(x) = \frac{2x-1}{x+2}.$$

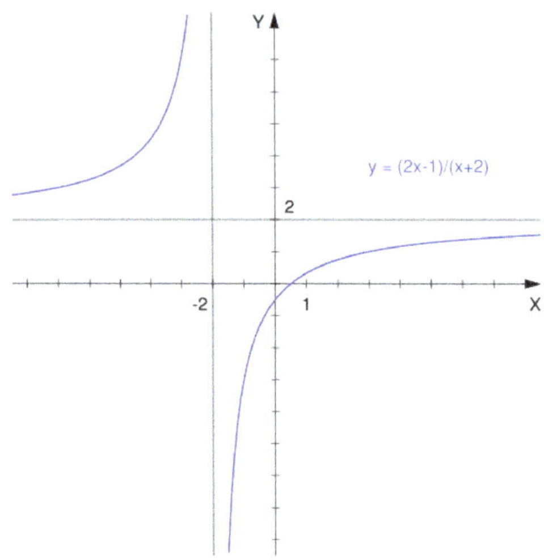

The graph of a continuous rational function. The function is not defined for $x=-2$. The vertical and horizontal lines are asymptotes.

is defined for all real numbers $x \neq -2$ and is continuous at every such point. Thus it is a continuous function. The question of continuity at $x = -2$ does not arise, since $x = -2$ is not in the domain of y. There is no continuous function F: R → R that agrees with $y(x)$ for all $x \neq -2$.

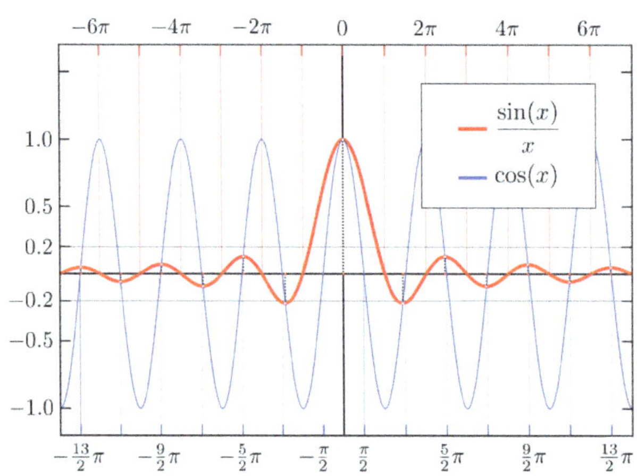

The sinc and the cos functions

Since the sine-function is continuous on all reals, the sinc function $G(x) = \sin x/x$, when defined for all real $x \neq 0$, is continuous at these points. Thus G is a continuous function there, too. However, unlike the previous example, G *can* be extended to a continuous function on *all* real numbers, by *defining* the value $G(0)$ to be the limit of $G(x)$, when x approaches 0, i.e.,

$$G(0) = \lim_{x \to 0} \frac{\sin x}{x} = 1$$

and thus

$$G(x) = \begin{cases} \dfrac{\sin(x)}{x} & if\, x \neq 0 \\ 1 & if\, x = 0, \end{cases}$$

the sinc-function becomes a continuous function on all real numbers. The term removable singularity is used in such cases, when (re)defining values of a function to coincide with the appropriate limits make a function continuous at specific points.

A more involved construction of continuous functions is the function composition. Given two continuous functions

$$f : D_f\ (\subseteq \mathbf{R}) \to R_f\ (\subseteq \mathbf{R}) \text{ and } g : D_g\ (R_f \subseteq D_g) \to R_g\ (\subseteq \mathbf{R}),$$

their composition being denoted as

$$c = g \circ f : D_f \to \mathbf{R}, \text{ and defined by } c(x) = g(f(x)), \text{is continuous.}$$

This construction allows stating, for example, that

$e^{\sin(\ln x)}$ is continuous for all $x > 0$.

Examples of Discontinuous Functions

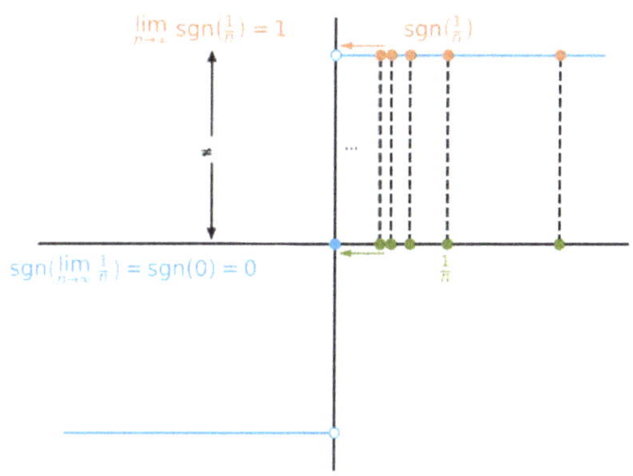

Plot of the signum function

It shows that $\lim_{n\to\infty} \operatorname{sgn}\left(\frac{1}{n}\right) \neq \operatorname{sgn}\left(\lim_{n\to\infty} \frac{1}{n}\right)$. Thus, the signum function is discontinuous at 0 for point 2.1.3.

An example of a discontinuous function is the Heaviside step function H, defined by

$$H(x) = \begin{cases} 1 & if\, x \geq 0 \\ 0 & if\, x < 0 \end{cases}$$

Pick for instance $\varepsilon = 1/2$. Then there is no δ-neighborhood around $x = 0$, i.e. no open interval $(-\delta, \delta)$ with $\delta > 0$, that will force all the $H(x)$ values to be within the ε-neighborhood of $H(0)$, i.e. within $(1/2, 3/2)$. Intuitively we can think of this type of discontinuity as a sudden jump in function values.

Similarly, the signum or sign function

$$\operatorname{sgn}(x) = \begin{cases} 1 & if\, x > 0 \\ 0 & if\, x = 0 \\ -1 & if\, x < 0 \end{cases}$$

is discontinuous at $x = 0$ but continuous everywhere else. Yet another example: the function

$$f(x) = \begin{cases} \sin\left(x^{-2}\right) & if\, x \neq 0 \\ 0 & if\, x = 0 \end{cases}$$

is continuous everywhere apart from $x = 0$

Besides plausible continuities and discontinuities like above, there are also functions with a behavior, often coined pathological, for example, Thomae's function,

$$f(x) = \begin{cases} 1 & if\, x = 0 \\ \frac{1}{q} & if\, x = \frac{p}{q} \text{(in lowest terms) is a rational number} \\ 0 & if\, x \text{ is irrational.} \end{cases}$$

is continuous at all irrational numbers and discontinuous at all rational numbers. In a similar vein, Dirichlet's function, the indicator function for the set of rational numbers,

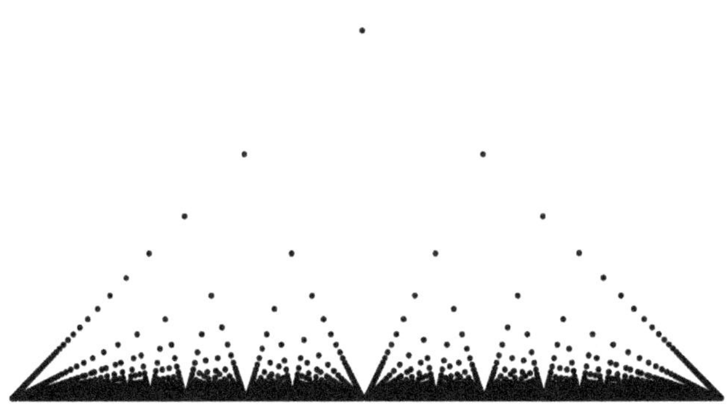

Point plot of Thomae's function on the interval (0,1). The topmost point in the middle shows f(1/2) = 1/2.

$$D(x) = \begin{cases} 0 & if\, x \text{ is irrational } (\in \mathbb{R} \setminus \mathbb{Q}) \\ 1 & if\, x \text{ is rational } (\in \mathbb{Q}) \end{cases}$$

is nowhere continuous.

Properties

Intermediate Value Theorem

The intermediate value theorem is an existence theorem, based on the real number property of completeness, and states:

> If the real-valued function f is continuous on the closed interval $[a, b]$ and k is some number between $f(a)$ and $f(b)$, then there is some number c in $[a, b]$ such that $f(c) = k$.

For example, if a child grows from 1 m to 1.5 m between the ages of two and six years, then, at some time between two and six years of age, the child's height must have been 1.25 m.

As a consequence, if f is continuous on $[a, b]$ and $f(a)$ and $f(b)$ differ in sign, then, at some point c in $[a, b]$, $f(c)$ must equal zero.

Extreme Value Theorem

The extreme value theorem states that if a function f is defined on a closed interval $[a,b]$ (or any closed and bounded set) and is continuous there, then the function attains its maximum, i.e. there exists $c \in [a,b]$ with $f(c) \geq f(x)$ for all $x \in [a,b]$. The same is true of the minimum of f. These statements are not, in general, true if the function is defined on an open interval (a,b) (or any set that is not both closed and bounded), as, for example, the continuous function $f(x) = 1/x$, defined on the open interval (0,1), does not attain a maximum, being unbounded above.

Relation to Differentiability and Integrability

Every differentiable function

$$f : (a,b) \to \mathbf{R}$$

is continuous, as can be shown. The converse does not hold: for example, the absolute value function

$$f(x) = |x| = \begin{cases} x & if\, x \geq 0 \\ -x & if\, x < 0 \end{cases}$$

is everywhere continuous. However, it is not differentiable at $x = 0$ (but is so everywhere else). Weierstrass's function is also everywhere continuous but nowhere differentiable.

The derivative $f'(x)$ of a differentiable function $f(x)$ need not be continuous. If $f'(x)$ is continuous, $f(x)$ is said to be continuously differentiable. The set of such functions is denoted $C^1((a,b))$. More generally, the set of functions

$$f : \Omega \to \mathbf{R}$$

(from an open interval (or open subset of R) Ω to the reals) such that f is n times differentiable and such that the n-th derivative of f is continuous is denoted $C^n(\Omega)$. In the field of computer graphics, these three levels are sometimes called G^0 (continuity of position), G^1 (continuity of tangency), and G^2 (continuity of curvature).

Every continuous function

$$f:[a,b]\to \mathbf{R}$$

is integrable (for example in the sense of the Riemann integral). The converse does not hold, as the (integrable, but discontinuous) sign function shows.

Pointwise and Uniform Limits

Given a sequence

$$f_1, f_2, \ldots: I \to \mathbf{R}$$

of functions such that the limit

$$f(x) := \lim_{n\to\infty} f_n(x)$$

exists for all x in D, the resulting function $f(x)$ is referred to as the pointwise limit of the sequence of functions $(f_n)_{n\in N}$. The pointwise limit function need not be continuous, even if all functions f_n are continuous. However, f is continuous if all functions f_n are continuous and the sequence converges uniformly, by the uniform convergence theorem. This theorem can be used to show that the exponential functions, logarithms, square root function, trigonometric functions are continuous.

Directional and Semi-continuity

Discontinuous functions may be discontinuous in a restricted way, giving rise to the concept of directional continuity (or right and left continuous functions) and semi-continuity. Roughly speaking, a function is *right-continuous* if no jump occurs when the limit point is approached from the right. Formally, f is said to be right-continuous at the point c if the following holds: For any number $\epsilon > 0$ however small, there exists some number $\delta > 0$ such that for all x in the domain with $c < x < c + \delta$, the value of $f(x)$ will satisfy

$$|f(x) - f(c)| < \varepsilon.$$

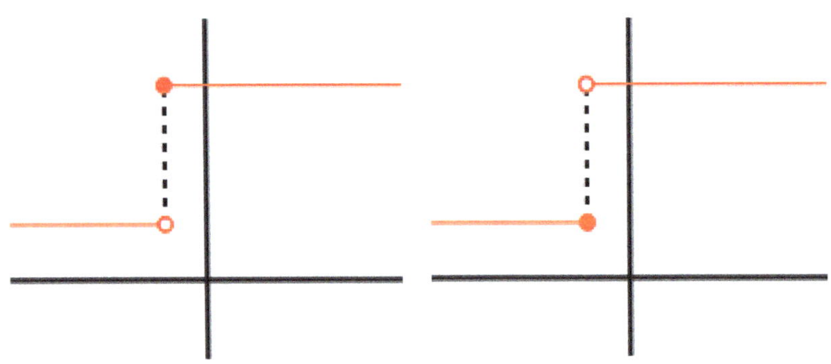

A right-continuous function A left-continuous function

This is the same condition as for continuous functions, except that it is required to hold for x strictly larger than c only. Requiring it instead for all x with $c - \delta < x < c$ yields the notion of *left-continuous* functions. A function is continuous if and only if it is both right-continuous and left-continuous.

A function f is *lower semi-continuous* if, roughly, any jumps that might occur only go down, but not up. That is, for any $\epsilon > 0$, there exists some number $\delta > 0$ such that for all x in the domain with $|x - c| < \delta$, the value of $f(x)$ satisfies

$$f(x) \geq f(c) - \epsilon.$$

The reverse condition is *upper semi-continuity*.

Continuous Functions between Metric Spaces

The concept of continuous real-valued functions can be generalized to functions between metric spaces. A metric space is a set X equipped with a function (called metric) d_X, that can be thought of as a measurement of the distance of any two elements in X. Formally, the metric is a function

$$d_X : X \times X \to \mathbf{R}$$

that satisfies a number of requirements, notably the triangle inequality. Given two metric spaces (X, d_X) and (Y, d_Y) and a function

$$f : X \to Y$$

then f is continuous at the point c in X (with respect to the given metrics) if for any positive real number ϵ, there exists a positive real number δ such that all x in X satisfying $d_X(x, c) < \delta$ will also satisfy $d_Y(f(x), f(c)) < \epsilon$. As in the case of real functions above, this is equivalent to the condition that for every sequence (x_n) in X with limit $\lim x_n = c$, we have $\lim f(x_n) = f(c)$. The latter condition can be weakened as follows: f is continuous at the point c if and only if for every convergent sequence (x_n) in X with limit c, the sequence $(f(x_n))$ is a Cauchy sequence, and c is in the domain of f.

The set of points at which a function between metric spaces is continuous is a G_δ set – this follows from the ϵ-δ definition of continuity.

This notion of continuity is applied, for example, in functional analysis. A key statement in this area says that a linear operator

$$T : V \to W$$

between normed vector spaces V and W (which are vector spaces equipped with a compatible norm, denoted $||x||$) is continuous if and only if it is bounded, that is, there is a constant K such that

$$\|T(x)\| \leq K\|x\|$$

for all x in V.

Uniform, Hölder and Lipschitz Continuity

The concept of continuity for functions between metric spaces can be strengthened in various ways by limiting the way δ depends on ϵ and c in the definition above. Intuitively, a function f as above is uniformly continuous if the δ does not depend on the point c. More precisely, it is required that for every real number $\epsilon > 0$ there exists $\delta > 0$ such that for every $c, b \in X$ with $d_X(b, c) < \delta$, we have that $d_Y(f(b), f(c)) < \epsilon$. Thus, any uniformly continuous function is continuous. The converse does not hold in general, but holds when the domain space X is compact. Uniformly continuous maps can be defined in the more general situation of uniform spaces.

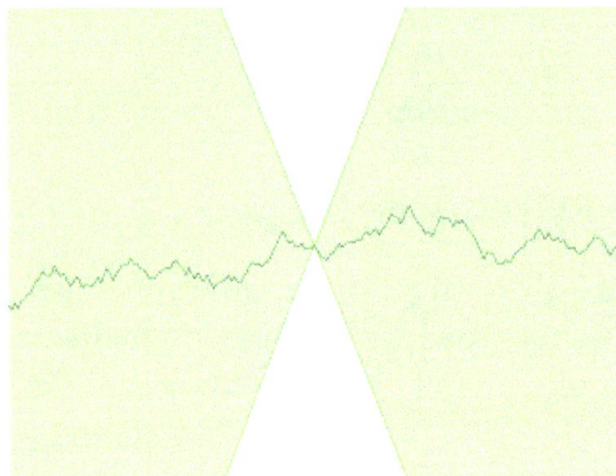

For a Lipschitz continuous function, there is a double cone (shown in white) whose vertex can be translated along the graph, so that the graph always remains entirely outside the cone.

A function is Hölder continuous with exponent α (a real number) if there is a constant K such that for all b and c in X, the inequality

$$d_Y(f(b), f(c)) \leq K \cdot (d_X(b,c))^{\alpha}$$

holds. Any Hölder continuous function is uniformly continuous. The particular case $\alpha = 1$ is referred to as Lipschitz continuity. That is, a function is Lipschitz continuous if there is a constant K such that the inequality

$$d_Y(f(b), f(c)) \leq K \cdot d_X(b,c)$$

holds for any b, c in X. The Lipschitz condition occurs, for example, in the Picard–Lindelöf theorem concerning the solutions of ordinary differential equations.

Continuous Functions between Topological Spaces

Another, more abstract, notion of continuity is continuity of functions between topological spaces in which there generally is no formal notion of distance, as there is in the case of metric spaces. A topological space is a set X together with a topology on X, which is a set of subsets of X satisfying a few requirements with respect to their unions and intersections that generalize the properties of the open balls in metric spaces while still allowing to talk about the neighbourhoods of a given point. The elements of a topology are called open subsets of X (with respect to the topology).

A function

$$f : X \to Y$$

between two topological spaces X and Y is continuous if for every open set $V \subseteq Y$, the inverse image

$$f^{-1}(V) = \{x \in X \mid f(x) \in V\}$$

is an open subset of X. That is, f is a function between the sets X and Y (not on the elements of the topology T_X), but the continuity of f depends on the topologies used on X and Y.

This is equivalent to the condition that the preimages of the closed sets (which are the complements of the open subsets) in Y are closed in X.

An extreme example: if a set X is given the discrete topology (in which every subset is open), all functions

$$f : X \to T$$

to any topological space T are continuous. On the other hand, if X is equipped with the indiscrete topology (in which the only open subsets are the empty set and X) and the space T set is at least T_0, then the only continuous functions are the constant functions. Conversely, any function whose range is indiscrete is continuous.

Continuity at a Point

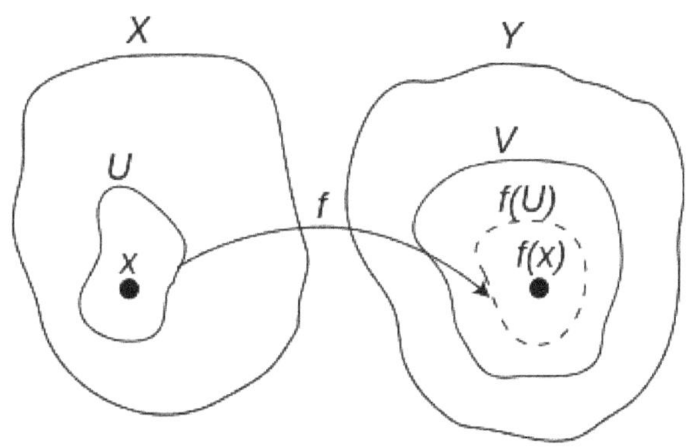

Continuity at a point: For every neighborhood V of $f(x)$, there is a neighborhood U of x such that $f(U) \subseteq V$

The translation in the language of neighborhoods of the (ϵ, δ)-definition of continuity leads to the following definition of the continuity at a point:

A function $f : X \to Y$ is continuous at a point $x \in X$ if and only if for any neighborhood V of $f(x)$ in Y, there is a neighborhood U of x such that $f(U) \subseteq V$.

This definition is equivalent to the same statement with neighborhoods restricted to open neighborhoods and can be restated in several ways by using preimages rather than images.

Also, as every set that contains a neighborhood is also a neighborhood, and $f^{-1}(V)$ is the largest subset U of X such that $f(U) \subseteq V$, this definition may be simplified into:

A function $f : X \to Y$ is continuous at a point $x \in X$ if and only if $f^{-1}(V)$ is a neighborhood of x for every neighborhood V of $f(x)$ in Y.

As an open set is a set that is a neighborhood of all its points, a function $f : X \to Y$ is continuous at every point of X if and only if it is a continuous function.

If X and Y are metric spaces, it is equivalent to consider the neighborhood system of open balls centered at x and $f(x)$ instead of all neighborhoods. This gives back the above δ-ϵ definition of continuity in the context of metric spaces. In general topological spaces, there is no notion of nearness or distance. If however the target space is a Hausdorff space, it is still true that f is continuous at a if and only if the limit of f as x approaches a is $f(a)$. At an isolated point, every function is continuous.

Alternative Definitions

Several equivalent definitions for a topological structure exist and thus there are several equivalent ways to define a continuous function.

Sequences and Nets

In several contexts, the topology of a space is conveniently specified in terms of limit points. In many instances, this is accomplished by specifying when a point is the limit of a sequence, but for some spaces that are too large in some sense, one specifies also when a point is the limit of more general sets of points indexed by a directed set, known as nets. A function is (Heine-)continuous only if it takes limits of sequences to limits of sequences. In the former case, preservation of limits is also sufficient; in the latter, a function may preserve all limits of sequences yet still fail to be continuous, and preservation of nets is a necessary and sufficient condition.

In detail, a function $f: X \to Y$ is sequentially continuous if whenever a sequence (x_n) in X converges to a limit x, the sequence $(f(x_n))$ converges to $f(x)$. Thus sequentially continuous functions "preserve sequential limits". Every continuous function is sequentially continuous. If X is a first-countable space and countable choice holds, then the converse also holds: any function preserving sequential limits is continuous. In particular, if X is a metric space, sequential continuity and continuity are equivalent. For non first-countable spaces, sequential continuity might be strictly weaker than continuity. (The spaces for which the two properties are equivalent are called sequential spaces.) This motivates the consideration of nets instead of sequences in general topological spaces. Continuous functions preserve limits of nets, and in fact this property characterizes continuous functions.

Closure Operator Definition

Instead of specifying the open subsets of a topological space, the topology can also be determined by a closure operator (denoted cl) which assigns to any subset $A \subseteq X$ its closure, or an interior operator (denoted int), which assigns to any subset A of X its interior. In these terms, a function

$$f:(X,cl) \to (X',cl')$$

between topological spaces is continuous in the sense above if and only if for all subsets A of X

$$f(cl(A)) \subseteq cl'(f(A)).$$

That is to say, given any element x of X that is in the closure of any subset A, $f(x)$ belongs to the closure of $f(A)$. This is equivalent to the requirement that for all subsets A' of X'

$$f^{-1}(cl'(A')) \supseteq cl(f^{-1}(A')).$$

Moreover,

$$f:(X,int) \to (X',int')$$

is continuous if and only if

$$f^{-1}(int'(A')) \subseteq int(f^{-1}(A'))$$

for any subset A' of Y.

Properties

If $f: X \to Y$ and $g: Y \to Z$ are continuous, then so is the composition $g \circ f: X \to Z$. If $f: X \to Y$ is continuous and

- X is compact, then $f(X)$ is compact.
- X is connected, then $f(X)$ is connected.
- X is path-connected, then $f(X)$ is path-connected.
- X is Lindelöf, then $f(X)$ is Lindelöf.
- X is separable, then $f(X)$ is separable.

The possible topologies on a fixed set X are partially ordered: a topology τ_1 is said to be coarser than another topology τ_2 (notation: $\tau_1 \subseteq \tau_2$) if every open subset with respect to τ_1 is also open with respect to τ_2. Then, the identity map

$$\mathrm{id}_X: (X, \tau_2) \to (X, \tau_1)$$

is continuous if and only if $\tau_1 \subseteq \tau_2$. More generally, a continuous function

$$(X, \tau_X) \to (Y, \tau_Y)$$

stays continuous if the topology τ_Y is replaced by a coarser topology and/or τ_X is replaced by a finer topology.

Homeomorphisms

Symmetric to the concept of a continuous map is an open map, for which *images* of open sets are open. In fact, if an open map f has an inverse function, that inverse is continuous, and if a continuous map g has an inverse, that inverse is open. Given a bijective function f between two topological spaces, the inverse function f^{-1} need not be continuous. A bijective continuous function with continuous inverse function is called a *homeomorphism.*

If a continuous bijection has as its domain a compact space and its codomain is Hausdorff, then it is a homeomorphism.

Defining Topologies via Continuous Functions

Given a function

$$f: X \to S,$$

where X is a topological space and S is a set (without a specified topology), the final topology on S is defined by letting the open sets of S be those subsets A of S for which $f^{-1}(A)$ is open in X. If S has an existing topology, f is continuous with respect to this topology if and only if the existing topology is coarser than the final topology on S. Thus the final topology can be characterized as the finest topology on S that makes f continuous. If f is surjective, this topology is canonically identified with the quotient topology under the equivalence relation defined by f.

Dually, for a function f from a set S to a topological space, the initial topology on S has as open subsets A of S those subsets for which $f(A)$ is open in X. If S has an existing topology, f is continuous with respect to this topology if and only if the existing topology is finer than the initial topology on S. Thus the initial topology can be characterized as the coarsest topology on S that makes f continuous. If f is injective, this topology is canonically identified with the subspace topology of S, viewed as a subset of X.

A topology on a set S is uniquely determined by the class of all continuous functions $S \to X$ into all topological spaces X. Dually, a similar idea can be applied to maps $X \to S$.

Related Notions

Various other mathematical domains use the concept of continuity in different, but related meanings. For example, in order theory, an order-preserving function $f: X \to Y$ between particular types of partially ordered sets X and Y is continuous if for each directed subset A of X, we have $\sup(f(A)) = f(\sup(A))$. Here sup is the supremum with respect to the orderings in X and Y, respectively. This notion of continuity is the same as topological continuity when the partially ordered sets are given the Scott topology.

In category theory, a functor

$$F : \mathcal{C} \to \mathcal{D}$$

between two categories is called *continuous*, if it commutes with small limits. That is to say,

$$\varprojlim_{i \in I} F(C_i) \cong F\left(\varprojlim_{i \in I} C_i\right)$$

for any small (i.e., indexed by a set I, as opposed to a class) diagram of objects in $\mathcal{C}$.

A *continuity space* is a generalization of metric spaces and posets, which uses the concept of quantales, and that can be used to unify the notions of metric spaces and domains.

Fundamental Theorem of Calculus

It relates the derivative to the integraland provides the principal method for evaluating definite integrals. In brief, it states that any function that is continuous over an interval has an antiderivative (a function whose rate of change, or derivative, equals the function) on that interval. Further, the definite integral of such a function over an interval $a < x < b$ is the difference $F(b) - F(a)$, where F is an antiderivative of the function. This particularly elegant theorem shows the inverse function relationship of the derivative and the integral and serves as the backbone of the physical sciences. It was articulated independently by Isaac Newton and Gottfried Wilhelm Leibniz.

The First Fundamental Theorem of Calculus

We have learned about indefinite integrals, which was the process of finding the antiderivative of a function. In contrast to the indefinite integral, the result of a definite integral will be a number, instead of a function. The definite integral of a function is the signed area under the graph of the function, and is expressed in the form of $\int_a^b f(x)\,dx$:

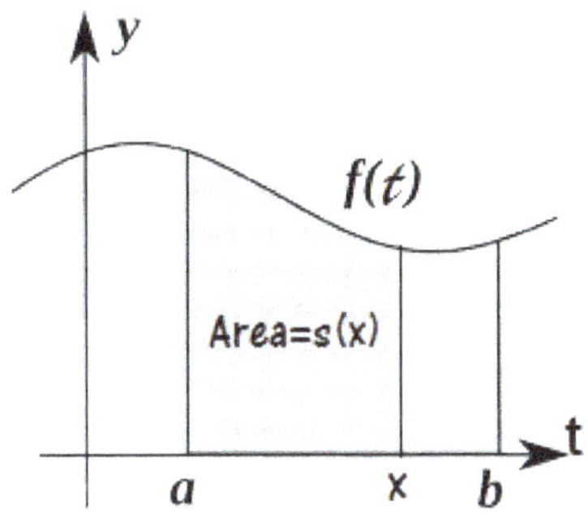

Now, suppose that we formed an area function $S(x)$ in such a way that it is dependent on the function $f(x)$ as

$$S(x) = \int_a^x f(t)\,dt,$$

where is f continuous on the interval $[a,b]$. Now, suppose we wanted to find the the rate of change of the area with respect to x:

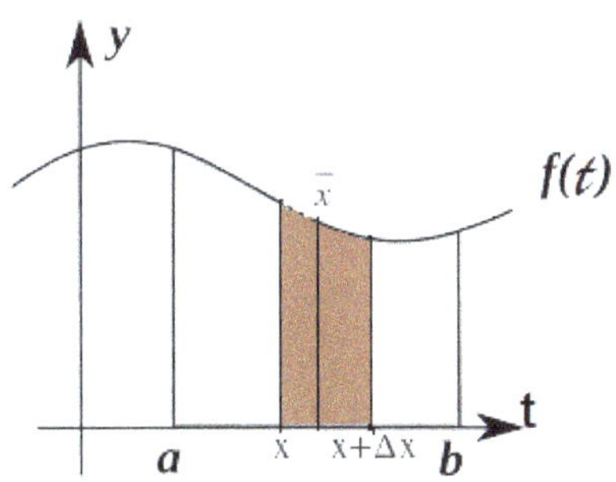

We can see from the figure above that the area of the shaded region is equal to the area under the curve $f(t)$ from a to $x+\Delta x$ minus the area under $f(t)$ from a to x. Thus,

$$\Delta S = A(x+\Delta x) - A(x)$$

$$\frac{\Delta S}{\Delta x} = \frac{A(x+\Delta x) - A(x)}{\Delta x}.$$

So, the rate of change of area becomes

$$S'(x) = \frac{dS}{dx} = \lim_{\Delta x \to 0} \frac{S(x+\Delta x) - S(x)}{\Delta x}.$$

We know that there is an $\overline{x}$ found between x and $x+\Delta x$ such that the area of the shaded region is equal to $f(\overline{x})\,\Delta x$:

$$\begin{aligned} S'(x) &= \lim_{\Delta x\to 0}\frac{S\left(x+\Delta x\right)-S(x)}{\Delta x} \\ &= \lim_{\Delta x\to 0}\frac{f(\overline{x})\,\Delta x}{\Delta x} \\ &= \lim_{\Delta x\to 0} f(\overline{x}) \\ &= f(x). \end{aligned}$$

The last step is true because, as $\Delta x \to 0$, anything found between x and $x+\Delta x$ approaches x. So, now we are ready to state the first fundamental theorem of calculus.

Theorem

If f is continuous on $[a,b]$, then the function defined by

$$S(x)=\int_a^x f(t)\,dt$$

is continuous on $[a,b]$ and differentiable on (a,b), and $S'(x)=f(x)$

So basically integration is the opposite of differentiation. More clearly, the first fundamental theorem of calculus can be rewritten in Leibniz notation as

$$\frac{d}{dx}\int_a^x f(t)\,dt = f(x).$$

Example

Find the derivative of $k(x)=\int_2^x \left(4^t+t\right)dt$.

The function f is continuous, so from the first fundamental theorem of calculus we have

$$k'(x)=4^x+x.$$

Example

What is the derivative of $h(x)=\int_2^{x^2}\frac{1}{1+t^2}\,dt$?

We use the first fundamental theorem of calculus in accordance with the chain-rule to solve this.

Let $u=x^2$, then

$$\frac{d}{dx}\int_2^{x^2}\frac{1}{1+t^2}\,dt=\frac{d}{du}\left[\int_1^u\frac{1}{1+t^2}\,dt\right]\cdot\frac{du}{dx}$$

$$= \frac{1}{1+u^2}.2x$$
$$= \frac{2x}{1+x^4}.$$

Example

Find the derivative of $\int_1^{x^2} \cos t \, dt$.

Again, we use the chain rule along with the fundamental theorem of calculus to solve this.

Let $u = x^2$, then

$$\begin{aligned}\int_1^{x^2} \cos t \, dt &= \frac{d}{du}\left[\int_1^u \cos t \, dt\right].\frac{du}{dx} \\ &= \cos u.\frac{d}{dx}\left(x^2\right) \\ &= \cos u.\, 2x \\ &= 2x\cos x^2.\end{aligned}$$

Example

Find the derivative of $h(x) = \int_x^{3x} \sin\theta \, d\theta$.

We use the following property of integrals:

$$\int_b^a f(x)dx = \int_b^0 f(x)dx + \int_0^a f(x)dx.$$

So,

$$\begin{aligned}h'(x) &= \frac{d}{dx}\int_x^{3x} \sin\theta \, d\theta \\ &= \frac{d}{dx}\int_x^0 \sin\theta \, d\theta + \frac{d}{dx}\int_0^{3x} \sin\theta \, d\theta \\ &= \frac{d}{dx} - \int_0^x \sin\theta \, d\theta + \frac{d}{dx}\int_0^{3x} \sin\theta \, d\theta \\ &= -\sin x + \frac{d}{du}\int_0^u \sin\theta \, d\theta.\frac{du}{dx} \\ &= -\sin x + 3\sin 3x.\end{aligned}$$

Second Fundamental Theorem of Calculus

Theorem

If f is a continuous function on (a,b), then

$$\int_a^b f(x)\,dx = F(b) - F(a),$$

where F is the anti-derivative of f, i.e. $F' = f$.

Proof

We know from the first fundamental theorem of calculus that if $S(x) = \int_a^x f(t)\,dt$, then S is the anti-derivative of f or $S'(x) = f(x)$ And since $F'(x) = f(x)$

$$S'(x) = F'(x)$$

Integrating both sides with respect to x, we have

$$S(x) = \int_a^x f(t)\,dt = F(x) + Cb,$$

Where C is some constant.

Now, plugging in $x = a$ in this equation, we have

$$S(a) = \int_a^a f(t)\,dt = F(a) + C.$$

By definition, $S(a) = \int_a^a f(t)\,dt = 0$, and hence

$$F(a) + C = 0$$
$$\Rightarrow C = -F(a).$$

Plugging in $C = -F(a)$ back in the equation (1), we have

$$S(x) = \int_a^x f(t)\,dt = F(x) - F(a).$$

Finally, setting $x = b$, we have

$$S(b) = \int_a^b f(t)\,dt = F(b) - F(a)$$
$$\Rightarrow \int_a^b f(x)\,dx = F(b) - F(a).$$

Proof

We could also prove the above theorem using the concept of Riemann sums.

We know from the first fundamental theorem of calculus that if $S(x) = \int_a^x f(t)\,dt$, then S is the anti-derivative of f or $S'(x) = f(x)$. So we know that $f(x) = F'(x)$

Consider a partition $P = \{a = x_0 < x_1 < x_2 < ... < x_n = b\}$. With this partition as reference, we can write.

$$F(b) - F(a) = \sum_{i=0}^{n-1} [F(x_i + 1) - F(x_i)]$$

This is now the neater part of the proof. $\forall i = 0,1,2,...,n-1 \, \exists t_i \in (x_i, x_i + 1)$ such that $F(x_i + 1) - F(x_i) = F'(t_i)(x_i + 1 - x_i) = f(t_i)(x_{i+1} - x_i)$. This is direct implication of the mean value theorem. So now what we have is

$$F(b) - F(a) = \sum_{i=0}^{n-1} f(t_i)(x_{i+1} - x_1); t_i \in (x_i, x_i + 1).$$

The right-hand side of the expression is nothing but the Riemann sum which will eventually converge to definite integral as the partition P gets finer and finer:

$$F(b) - F(a) = \int_a^b f(t)\, dt.$$

This theorem transforms the difficult problem of evaluating definite integrals by calculating limits of sums, into an easier problem of finding the anti-derivative. So for example if we are asked to compute the integral $\int_a^b f(x)\, dx$, we find the anti-derivative of $f(x)$ and compute their value at each end-point of the integral, and finally subtract them from each other.

Example

Evaluate $\int_0^1 x^2\, dx$.

According to the fundamental theorem of calculus, we have

$$\int_0^1 x^2\, dx = F(1) - F(0),$$

where $F(x)$ is an anti-derivative of x^2.

Indefinite integration of x^2 gives

$$\int x^2\, dx = \frac{1}{3}x^3 + C,$$

where C is the constant of integration.

Hence we have

$$F(1) - F(0) = \left(\frac{1}{3} \times 1^3 + C\right) - \left(\frac{1}{3} \times 0 + C\right) = \frac{1}{3}.$$

Observe that the constant of integration C can be ignored since it is to be eliminated regardless of its value.

Example

Find the area under the curve $y = x^3 + 1$ from $x = -1$ to $x = 1$

Since $x^3 + 1 \geq 0$ over the interval $[-1,1]$, the area under the curve $y = x^3 + 1$ from $x = -1$ to $x = 1$ is equal to $\int_{-1}^{1} (x^3 + 1)\, dx$. since $\frac{1}{4}x^4 + x$ is an anti-derivative of $x^3 + 1$, the area under the curve is

$$\int_{-1}^{1} (x^3 + 1)\, dx = \left[\frac{1}{4}x^4 + x\right]_{-1}^{1}$$
$$= \left(\frac{1}{4}.1^4 + 1\right) - \left(\frac{1}{4}.(-1\ )^4 + (-1\ )\right)$$
$$= \frac{5}{4} - \left(-\frac{3}{4}\right)$$
$$= 2.$$

Example

Evaluate $\int_{-3}^{2} (2x^2 - 3x + 4)\, dx$.

Since $\int (2x^2 - 3x + 4)\, dx = \frac{2}{3}x^3 - \frac{3}{2}x^2 + 4x + C$, we have

$$\int_{-3}^{2} (2x^2 - 3x + 4)\, dx = \left[\frac{2}{3}x^3 - \frac{3}{2}x^2 + 4x\right]_{-3}^{2}$$
$$= \left(\frac{2}{3}.2^3 - \frac{3}{2}.2^3 + 4.2\right) - \left(\frac{2}{3}.(-3)^3 - \frac{3}{2}.(-3)^2 + 4.(-3)\right)$$
$$= \frac{305}{6}.$$

Example

Evaluate $\int_{4}^{9} \sqrt{x}\, dx$.

We have

$$\int_{4}^{9} \sqrt{x}\, dx = \int_{4}^{9} x^{\frac{1}{2}} dx$$
$$= \left[\frac{2}{3}x^{\frac{3}{2}}\right]_{4}^{9}$$
$$= \frac{2}{3}.9^{\frac{3}{2}} - \frac{2}{3}.4^{\frac{3}{2}}$$
$$= \frac{38}{3}.$$

Example

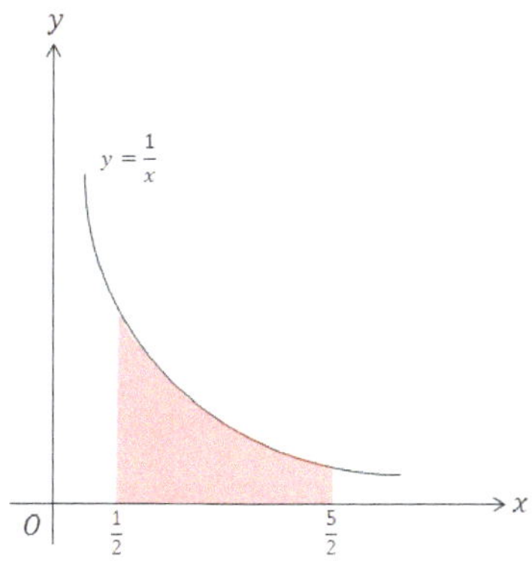

What is the area of the shaded region in the above figure?

We need to find the area under the curve $y = \frac{1}{x}$ from $x = \frac{1}{2}$ to $x = \frac{5}{2}$. since $\int \frac{1}{x}\,dx = \text{In } x + C$, the area under the curve is equal to

$$\int_{\frac{1}{2}}^{\frac{5}{2}} \frac{1}{x}\,dx = [\text{In } x]_{\frac{1}{2}}^{\frac{5}{2}}$$
$$= \text{In}\frac{5}{2} - \text{In}\frac{1}{2}$$
$$= \text{In } 5.$$

Example

Evaluate $\int_{-\frac{\pi}{4}}^{0} \sec x \tan x\,dx.$

We know that

$$\frac{d}{dx}\sec x = \sec x \tan x.$$

Thus, the integral would be

$$\int_{-\frac{\pi}{4}}^{0} \sec x \tan x\,dx = \sec x]^{0}_{-\frac{\pi}{4}}$$
$$= \sec 0 - \sec\left(\frac{-\pi}{4}\right)$$
$$= 1 - \sqrt{2}.$$

Example

Evaluate $\int_{-\frac{\pi}{2}}^{\frac{5}{6}} \sin x\,dx.$

We have

$$\int_{-\frac{\pi}{2}}^{\frac{5}{6}\pi} \sin x\, dx = \left[-\cos x\right]_{-\frac{\pi}{2}}^{\frac{5}{6}\pi}$$

$$= \sqrt{\frac{3}{2}}.$$

Example

Find the area under the curve $y = 2x - x^2$ from $x = 1$ to $x = 2$.

Since $2x - x^2 \geq 0$ over the interval $[1,2]$, the area under the curve $y = 2x - x^2$ from $x = 1$ to $x = 2$ is equal to $\int_1^2 \left(2x - x^2\right) dx$. since $x^2 - \frac{1}{3}x^3$ is an anti-derivative of $2x - x^2$,

$$\int_1^2 \left(2x - x^2\right) dx = \left[x^2 - \frac{1}{3}x^3\right]_1^2$$

$$= \left(2^2 - \frac{1}{3}.2^3\right) - \left(1^2 - \frac{1}{3}.1^3\right)$$

$$= \frac{2}{3}.$$

References

- Dun, Liu; Fan, Dainian; Cohen, Robert Sonné (1966). "A comparison of Archimdes' and Liu Hui's studies of circles". Chinese studies in the history and philosophy of science and technology. 130. Springer: 279. ISBN 0-7923-3463-9
- Felscher, Walter (2000), "Bolzano, Cauchy, Epsilon, Delta", American Mathematical Monthly, Mathematical Association of America, 107 (9): 844–862, doi:10.2307/2695743, JSTOR 2695743
- Zill, Dennis G.; Wright, Scott; Wright, Warren S. (2009). Calculus: Early Transcendentals (3 ed.). Jones & Bartlett Learning. p. xxvii. ISBN 0-7637-5995-3. Extract of page 27
- Page, Warren; Hersh, Reuben; Selden, Annie; et al., eds. (2002), "Media Highlights", The College Mathematics, Mathematical Association of America, 33 (2): 147–154, JSTOR Journal 2687124 Journal
- Hazewinkel, Michiel, ed. (2001) [1994], "Continuous function", Encyclopedia of Mathematics, Springer Science+Business Media B.V. / Kluwer Academic Publishers, ISBN 978-1-55608-010-4
- David Hestenes, Garrett Sobczyk: Clifford Algebra to Geometric Calculus, a Unified Language for mathematics and Physics (Dordrecht/Boston:G.Reidel Publ.Co., 1984, ISBN 90-277-2561-6
- Felscher, Walter (2000), "Bolzano, Cauchy, Epsilon, Delta", American Mathematical Monthly, Mathematical Association of America, 107 (9): 844–862, doi:10.2307/2695743, JSTOR 2695743

2

Differential Calculus

Differential calculus is an important sub-field of calculus, which studies the rates of change of quantities. The topics of derivatives and differential are integral to the study of differential calculus. This chapter discusses the fundamental concepts of differential calculus such as differentiation rules, differentiable functions and implicit functions.

A branch of mathematics dealing with the concepts of derivative and differential and the manner of using them in the study of functions. The development of differential calculus is closely connected with that of integral calculus. Indissoluble is also their content. Together they form the base of mathematical analysis, which is extremely important in the natural sciences and in technology. The introduction of variable magnitudes into mathematics by R. Descartes was the principal factor in the creation of differential calculus. Differential and integral calculus were created, in general terms, by I. Newton and G. Leibniz towards the end of the 17th century, but their justification by the concept of limit was only developed in the work of A.L. Cauchy in the early 19th century. The creation of differential and integral calculus initiated a period of rapid development in mathematics and in related applied disciplines. Differential calculus is usually understood to mean classical differential calculus, which deals with real-valued functions of one or more real variables, but its modern definition may also include differential calculus in abstract spaces. Differential calculus is based on the concepts of real number; function; limit and continuity – highly important mathematical concepts, which were formulated and assigned their modern content during the development of mathematical analysis and during studies of its foundations. The central concepts of differential calculus – the derivative and the differential – and the apparatus developed in this connection furnish tools for the study of functions which locally look like linear functions or polynomials, and it is in fact such functions which are of interest, more than other functions, in applications.

Derivative

Let a function $y = f(x)$ be defined in some neighbourhood of a point x_0. Let $\Delta x \neq 0$ denote the increment of the argument and let $\Delta y = f(x_0 + \Delta x) - f(x_0)$ denote the corresponding increment of the value of the function. If there exists a (finite or infinite) limit

$$\lim_{\Delta x \to 0} \frac{\Delta y}{\Delta x},$$

then this limit is said to be the derivative of the function f at x_0; it is denoted by $f'(x_0)$, $df(x_0)/dx$, y', $y'x$, dy/dx. Thus, by definition,

$$f'(x_0) = \lim_{\Delta x \to 0} \frac{\Delta y}{\Delta x} = \lim_{\Delta x \to 0} \frac{f(x_0 + \Delta x) - f(x_0)}{\Delta x}.$$

The operation of calculating the derivative is called differentiation. If $f'(x_0)$ is finite, the function f is called differentiable at the point x_0. A function which is differentiable at each point of some interval is called differentiable in the interval.

Geometric Interpretation of the Derivative

Let C be the plane curve defined in an orthogonal coordinate system by the equation $y = f(x)$ where f is defined and is continuous in some interval J; let $M(x_0, y_0)$ be a fixed point on C, let $P(x, y)$ $(x \in J)$ be an arbitrary point of the curve C and let MP be the secant (Figure). An oriented straight line MT (T a variable point with abscissa $x_0 + \Delta x$) is called the tangent to the curve C at the point M if the angle between the secant MP and the oriented straight line tends to zero as $x \to x_0$ (in other words, as the point $P \in C$ arbitrarily tends to the point M). If such a tangent exists, it is unique. Putting $x = x_0 + \Delta x$, $\Delta y = f(x_0 + \Delta x) - f(x_0)$, one obtains the equation $\tan \beta = \Delta y / \Delta x$ for the angle β between MP and the positive direction of the x-axis (Figure).

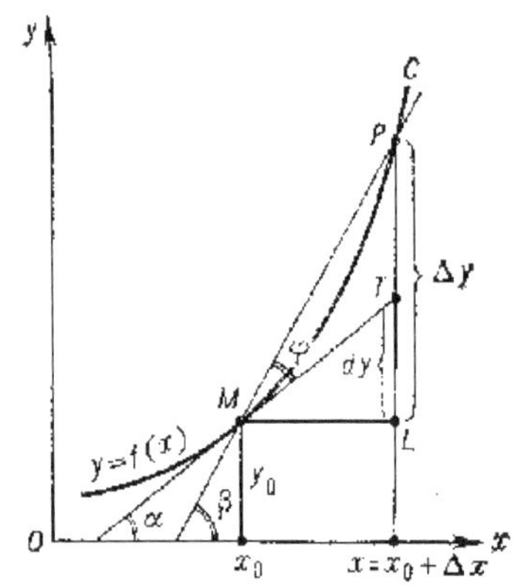

Figure

The curve C has a tangent at the point M if and only if $\lim_{\Delta x \to 0} \Delta y / \Delta x$ exists, i.e. if $f'(x_0)$ exists. The equation $\tan \alpha = f'(x_0)$ is valid for the angle α between the tangent and the positive direction of the x-axis. If $f'(x_0)$ is finite, the tangent forms an acute angle with the positive x-axis, i.e. $-\pi/2 < \alpha < \pi/2$; if $f'(x_0) = \infty$, the tangent forms a right angle with that axis (cf. Figure).

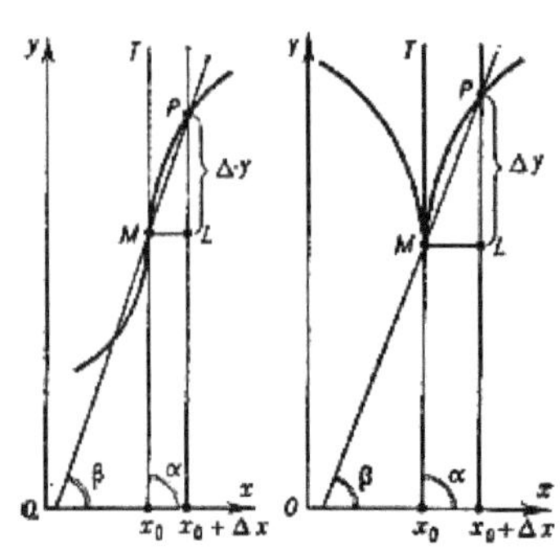

Figure

Thus, the derivative of a continuous function f at a point x_0 is identical to the slope $\tan\alpha$ of the tangent to the curve defined by the equation $y = f(x)$ at its point with abscissa x_0.

Mechanical Interpretation of the Derivative

Let a point M move in a straight line in accordance with the law $s = f(t)$. During time Δt the point M becomes displaced by $\Delta s = f(t+\Delta t) - f(t)$. The ratio $\Delta s / \Delta t$ represents the average velocity v_{av} during the time Δt. If the motion is non-uniform, v_{av} is not constant. The instantaneous velocity at the moment t is the limit of the average velocity as $\Delta t \to 0$, i.e. v= $f'(t)$ (on the assumption that this derivative in fact exists).

Thus, the concept of derivative constitutes the general solution of the problem of constructing tangents to plane curves, and of the problem of calculating the velocity of a rectilinear motion. These two problems served as the main motivation for formulating the concept of derivative.

A function which has a finite derivative at a point x_0 is continuous at this point. A continuous function need not have a finite nor an infinite derivative. There exist continuous functions having no derivative at any point of their domain of definition.

The formulas given below are valid for the derivatives of the fundamental elementary functions at any point of their domain of definition (exceptions are stated):

1) *if* $f(x) = C = \text{const}$, then $f'(x) = C' = 0$;

2) *if* $f(x) = x$, *then* $f'(x) = 1$

3) $(x^\alpha)'\alpha x^{\alpha-1}$, $\alpha = \text{const}$ ($x \neq 0$, *if* $\alpha \leq 1$);

4) $(\alpha^x)' = \alpha^x \text{ In } \alpha, \alpha = \text{const} > 0, \alpha \neq 1$; *in particular*, $(e^x)' = e^x$;

5) $(\log_\alpha x)' = (\log_\alpha e)/x = 1/(x \text{ In } \alpha)$, $\alpha = \text{const} > 0$, $\alpha \neq 1$, $(\text{In } x)' = 1/x$;

6) $(\sin x)' = \cos x$;

7) $(\cos x)' = -\sin x$;

8) $(\tan x)' = 1/\cos^2 x$;

9) $(\cot an\, x)' = -1/\sin^2 x$;

10) $(\arcsin x)' = 1/\sqrt{1-x^2}$, $x \neq \pm 1$;

11) $(\arccos x)' = -1/\sqrt{1-x^2}$, $x \neq \pm 1$;

12) $(\arctan x)' = 1/(1+x^2)$;

13) $(\text{arccotan } x)' = -1/(1+x^2)$;

14) $(\sinh x)' = \cosh x$;

15) $(\cosh x)' = \sinh x$;

16)$(\tanh x)' = 1/\cosh^2 x$;

17) $(\text{cot}\,\text{anh}\ x)' = -1/\sinh^2 x$.

The following laws of differentiation are valid:

If two functions u and v are differentiable at a point x_0, then the functions

$$cu \quad (where\ c = \text{const}), u \pm v, \qquad uv, \qquad \frac{u}{v}(v \neq 0)$$

are also differentiable at that point, and

$$(cu)' = cu',$$

$$(u \pm v)' = u' \pm v',$$

$$(uv)' = u'v + uv',$$

$$\left(\frac{u}{v}\right)' = \frac{u'v - uv'}{v^2}.$$

Theorem on the derivative of a composite function: If the function $y = f(u)$ is differentiable at a point u_0, while the function $\phi(x)$ is differentiable at a point x_0, and if $u_0 = \phi(x_0)$, then the composite function $y = f(\phi(x))$ is differentiable at x_0, and $y'x = f'(u_0)\phi'(x_0)$ or, using another notation, $dy/dx = (dy/du)(du/dx)$.

Theorem on the derivative of the inverse function: If $y = f(x)$ and $x = g(y)$ are two mutually inverse increasing (or decreasing) functions, defined on certain intervals, and if exists (i.e. is not infinite), then at the point $y_0 = f(x_0)$ the derivative $g'(y_0) = 1/f'(x_0)$ exists, or, in a different notation, $dx/dy = 1/(dy/dx)$. This theorem may be extended: If the other conditions hold and if also $f'(x_0) = 0$ or $f'(x_0) = \infty$, then, respectively, or $g'(y_0) = \infty\ or\ g'(y_0) = 0$.

One-sided Derivatives

If at a point (x_0) the limit

$$\lim_{\Delta x \downarrow 0} \frac{\Delta y}{\Delta x}$$

exists, it is called the right-hand derivative of the function $y = f(x)$ at (x_0) (in such a case the function need not be defined everywhere in a certain neighbourhood of the point (x_0); this requirement may then be restricted to $x \geq x_0$). The left-hand derivative is defined in the same way, as:

$$\lim_{\Delta x \uparrow 0} \frac{\Delta y}{\Delta x}$$

A function f has a derivative at a point (x_0) if and only if equal right-hand and left-hand derivatives exist at that point. If the function is continuous, the existence of a right-hand (left-hand) derivative at a point is equivalent to the existence, at the corresponding point of its graph, of a right (left) one-sided semi-tangent with slope equal to the value of this one-sided derivative. Points at which the semi-tangents do not form a straight line are called angular points or cusps (cf. Figure).

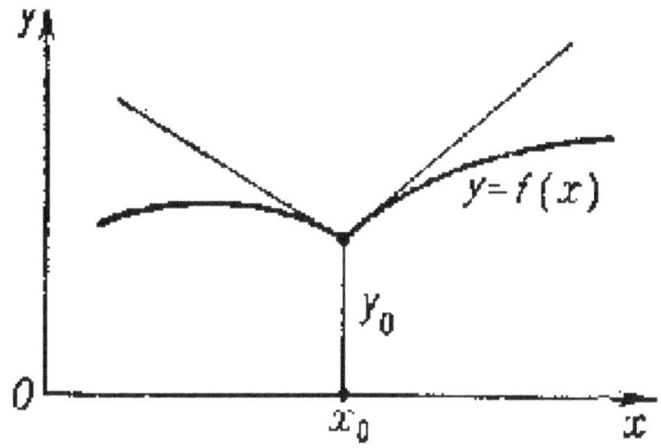

Figure

Derivatives of Higher Orders

Let a function $y = f(x)$ have a finite derivative $y' = f'(x)$ at all points of some interval; this derivative is also known as the first derivative, or the derivative of the first order, which, being a function of x, may in its turn have a derivative $y'' = f''(x)$, known as the second derivative, or the derivative of the second order, of the function f, etc. In general, the n -th derivative, or the derivative of order n , is defined by induction by the equation $y^{(n)} = (y^{(n-1)})'$, on the assumption that $y^{(n-1)}$ is defined on some interval. The notations employed along with $y^{(n)}$ are $f^{(n)}$, $d^n f(x)/dx^n$, and, if $n = 2,3$, also y'', $f''(x)$, y''', $f'''(x)$.

The second derivative has a mechanical interpretation: It is the acceleration $w = d^2s/dt^2 = f''(t)$ of a point in rectilinear motion according to the law $s = f(t)$.

Applications of Derivatives

Optimization

If f is a differentiable function on R (or an open interval) and x is a local maximum or a local minimum of f, then the derivative of f at x is zero. Points where $f'(x) = 0$ are called *critical points* or *stationary points* (and the value of f at x is called a *critical value*). If f is not assumed to be everywhere differentiable, then points at which it fails to be differentiable are also designated critical points.

If f is twice differentiable, then conversely, a critical point x of f can be analysed by considering the second derivative of f at x:

- if it is positive, x is a local minimum;
- if it is negative, x is a local maximum;

- if it is zero, then x could be a local minimum, a local maximum, or neither. (For example, $f(x) = x^3$ has a critical point at $x = 0$, but it has neither a maximum nor a minimum there, whereas $f(x) = \pm x^4$ has a critical point at $x = 0$ and a minimum and a maximum, respectively, there.)

This is called the second derivative test. An alternative approach, called the first derivative test, involves considering the sign of the f' on each side of the critical point.

Taking derivatives and solving for critical points is therefore often a simple way to find local minima or maxima, which can be useful in optimization. By the extreme value theorem, a continuous function on a closed interval must attain its minimum and maximum values at least once. If the function is differentiable, the minima and maxima can only occur at critical points or endpoints.

This also has applications in graph sketching: once the local minima and maxima of a differentiable function have been found, a rough plot of the graph can be obtained from the observation that it will be either increasing or decreasing between critical points.

In higher dimensions, a critical point of a scalar valued function is a point at which the gradient is zero. The second derivativetest can still be used to analyse critical points by considering the eigenvalues of the Hessian matrix of second partial derivatives of the function at the critical point. If all of the eigenvalues are positive, then the point is a local minimum; if all are negative, it is a local maximum. If there are some positive and some negative eigenvalues, then the critical point is called a "saddle point", and if none of these cases hold (i.e., some of the eigenvalues are zero) then the test is considered to be inconclusive.

Calculus of Variations

One example of an optimization problem is: Find the shortest curve between two points on a surface, assuming that the curve must also lie on the surface. If the surface is a plane, then the shortest curve is a line. But if the surface is, for example, egg-shaped, then the shortest path is not immediately clear. These paths are called geodesics, and one of the simplest problems in the calculus of variations is finding geodesics. Another example is: Find the smallest area surface filling in a closed curve in space. This surface is called a minimal surface and it, too, can be found using the calculus of variations.

Physics

Calculus is of vital importance in physics: many physical processes are described by equations involving derivatives, called differential equations. Physics is particularly concerned with the way quantities change and develop over time, and the concept of the "time derivative" — the rate of change over time — is essential for the precise definition of several important concepts. In particular, the time derivatives of an object's position are significant in Newtonian physics:

- velocity is the derivative (with respect to time) of an object's displacement (distance from the original position)
- acceleration is the derivative (with respect to time) of an object's velocity, that is, the second derivative (with respect to time) of an object›s position.

For example, if an object's position on a line is given by

$$x(t) = -16t^2 + 16t + 32,$$

then the object's velocity is

$$\dot{x}(t) = x'(t) = -32t + 16,$$

and the object's acceleration is

$$\ddot{x}(t) = x''(t) = -32,$$

which is constant.

Differential

Let a function $y = f(x)$ be defined in some neighbourhood of a point x and let there exist a number A such that the increment Δy may be represented as $\Delta y = A\Delta x + \omega$ with $\omega / \Delta x \to 0$ as $\Delta x \to 0$. The term $A \Delta x$ in this sum is denoted by the symbol dy or df and is named the differential of the function $f(x)$ (with respect to the variable x) at x. The differential is the principal linear part of increment of the function (its geometrical expression is the segment LT in Figure, where MT is the tangent to $y = f(x)$ at the point (x_0, y_0) under consideration).

The function $y = f(x)$ has a differential at x if and only if it has a finite derivative

$$f'(x) = \lim_{\Delta x \to 0} \frac{\Delta y}{\Delta x} = A$$

at this point. A function for which a differential exists is called differentiable at the point in question. Thus, the differentiability of a function implies the existence of both the differential and the finite derivative, and $dy = df(x) f'(x) \Delta x$. For the independent variable x one puts $dx = \Delta x$, and one may accordingly write $dy = f'(x)\, dx$, i.e. the derivative is equal to the ratio of the differentials:

$$f'(x) = \frac{dy}{dx}.$$

The formulas and the rules for computing derivatives lead to corresponding formulas and rules for calculating differentials. In particular, the theorem on the differential of a composite function is valid: If a function $y = f(u)$ is differentiable at a point u_0, while a function $\phi(x)$ is differentiable at a point x_0 and $u_0 = \phi(x_0)$, then the composite function $y = f(\phi(x_0))$ is differentiable at the point x_0 and $dy = f'(u_0)\, du$, where $du = \phi'(x_0)\, dx$. The differential of a composite function has exactly the form it would have if the variable u were an independent variable. This property is known the invariance of the form of the differential. However, if u is an independent variable, $du = \Delta u$ is an arbitrary increment, but if u is a function, du is the differential of this function which, in general, is not identical with its increment.

Differentials of Higher Orders

The differential dy is also known as the first differential, or differential of the first order. Let $y = f(x)$ have a differential $dy = f'(x)\, dx$ at each point of some interval. Here $dx = \Delta x$ is some number independent of x and one may say, therefore, that $dx = \text{const}$. The differential dy is a function of x alone, and may in turn have a differential, known as the second differential, or the differential of the second order, of f, etc. In general, the n-th differential, or the differential of order n, is defined by induction by the equality $d^n y = d(d^{n-1} y)$, on the assumption that the differential $d^{n-1} y$ is defined on some interval and that the value of dx is identical at all steps. The invariance condition for $d^2 y, d^3 y, \ldots$, is generally not satisfied (with the exception $y = f(u)$ where u is a linear function).

The repeated differential of dy has the form

$$\delta(dy) = f''(x)\, dx\, \delta x$$

and the value of $\delta(dy)$ for $dx = \delta x$ is the second differential.

Differential Equations

A differential equation is a relation between a collection of functions and their derivatives. An ordinary differential equation is a differential equation that relates functions of one variable to their derivatives with respect to that variable. A partial differential equation is a differential equation that relates functions of more than one variable to their partial derivatives. Differential equations arise naturally in the physical sciences, in mathematical modelling, and within mathematics itself. For example, Newton's second law, which describes the relationship between acceleration and force, can be stated as the ordinary differential equation

$$F(t) = m\frac{d^2x}{dt^2}.$$

The heat equation in one space variable, which describes how heat diffuses through a straight rod, is the partial differential equation

$$\frac{\partial u}{\partial t} = \alpha\frac{\partial^2 u}{\partial x^2}.$$

Here $u(x,t)$ is the temperature of the rod at position x and time t and α is a constant that depends on how fast heat diffuses through the rod.

Mean Value Theorem

The mean value theorem gives a relationship between values of the derivative and values of the original function. If $f(x)$ is a real-valued function and a and b are numbers with $a < b$, then the mean value theorem says that under mild hypotheses, the slope between the two points $(a, f(a))$ and $(b, f(b))$ is equal to the slope of the tangent line to f at some point c between a and b. In other words,

$$f'(c) = \frac{f(b) - f(a)}{b - a}$$

In practice, what the mean value theorem does is control a function in terms of its derivative. For instance, suppose that *f* has derivative equal to zero at each point. This means that its tangent line is horizontal at every point, so the function should also be horizontal. The mean value theorem proves that this must be true: The slope between any two points on the graph of *f* must equal the slope of one of the tangent lines of *f*. All of those slopes are zero, so any line from one point on the graph to another point will also have slope zero. But that says that the function does not move up or down, so it must be a horizontal line. More complicated conditions on the derivative lead to less precise but still highly useful information about the original function.

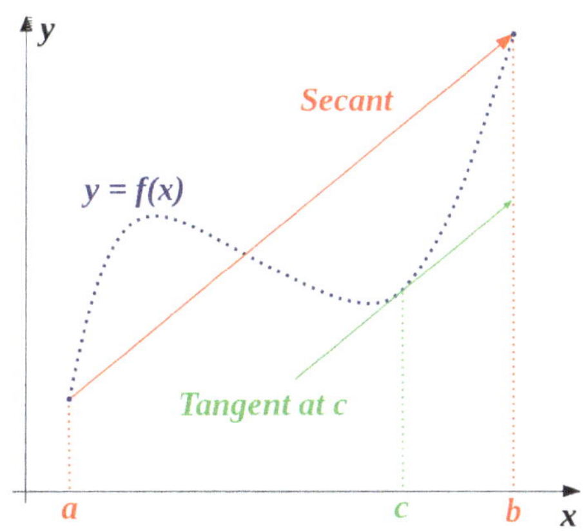

The mean value theorem: For each differentiable function $f:[a,b] \to \mathbb{R}$ with $a < b$ there is a $c \in (a,b)$ with $f'(c) = \frac{f(b)-f(a)}{b-a}$.

Taylor Polynomials and Taylor Series

The derivative gives the best possible linear approximation of a function at a given point, but this can be very different from the original function. One way of improving the approximation is to take a quadratic approximation. That is to say, the linearization of a real-valued function $f(x)$ at the point x_0 is a linear polynomial $a + b(x - x_0)$, and it may be possible to get a better approximation by considering a quadratic polynomial $a + b(x - x_0) + c(x - x_0)^2$. Still better might be a cubic polynomial $a + b(x - x_0) + c(x - x_0)^2 + d(x - x_0)^3$, and this idea can be extended to arbitrarily high degree polynomials. For each one of these polynomials, there should be a best possible choice of coefficients *a*, *b*, *c*, and *d* that makes the approximation as good as possible.

In the neighbourhood of x_0, for *a* the best possible choice is always $f(x_0)$, and for *b* the best possible choice is always $f'(x_0)$. For *c*, *d*, and higher-degree coefficients, these coefficients are determined by higher derivatives of *f*. *c* should always be $f''\frac{(x_0)}{2}$, and *d* should always be $f'''\frac{(x_0)}{3!}$. Using these coefficients gives the Taylor polynomial of *f*. The Taylor polynomial of degree *d* is the polynomial of degree *d* which best approximates *f*, and its coefficients can be found by a generalization of the above formulas. Taylor's theorem gives a precise bound on how good the approximation is. If *f* is a polynomial of degree less than or equal to *d*, then the Taylor polynomial of degree *d* equals *f*.

The limit of the Taylor polynomials is an infinite series called the Taylor series. The Taylor series is frequently a very good approximation to the original function. Functions which are equal to their Taylor series are called analytic functions. It is impossible for functions with discontinuities or sharp corners to be analytic, but there are smooth functions which are not analytic.

Implicit Function Theorem

Some natural geometric shapes, such as circles, cannot be drawn as the graph of a function. For instance, if $f(x, y) = x^2 + y^2 - 1$, then the circle is the set of all pairs (x, y) such that $f(x, y) = 0$. This set is called the zero set of f. It is not the same as the graph of f, which is a paraboloid. The implicit function theorem converts relations such as $f(x, y) = 0$ into functions. It states that if f is continuously differentiable, then around most points, the zero set of f looks like graphs of functions pasted together. The points where this is not true are determined by a condition on the derivative of f. The circle, for instance, can be pasted together from the graphs of the two functions $\pm\sqrt{1-x^2}$. In a neighborhood of every point on the circle except $(-1, 0)$ and $(1, 0)$, one of these two functions has a graph that looks like the circle. (These two functions also happen to meet $(-1, 0)$ and $(1, 0)$, but this is not guaranteed by the implicit function theorem.)

The implicit function theorem is closely related to the inverse function theorem, which states when a function looks like graphs of invertible functions pasted together.

Differential Calculus of Functions in Several Variables

For the sake of simplicity the case of functions in two variables (with certain exceptions) is considered below, but all relevant concepts are readily extended to functions in three or more variables. Let a function $z = f(x, y)$ be given in a certain neighbourhood of a point (x_0, y_0) and let the value $y = y_0$ be fixed. $f(x, y_0)$ will then be a function of x alone. If it has a derivative with respect to x at x_0, this derivative is called the partial derivative of f with respect to x at (x_0, y_0); it is denoted by $f'x(x_0, y_0)$. $\partial f(x_0, y_0)/\partial x$, $\partial f/\partial x$, $z'x$, $\partial z/\partial x$, or $f'_x(x_0, y_0)$. Thus, by definition,

$$f'_x(x_0, y_0) = \lim_{\Delta x \to 0} \frac{\Delta_x z}{\Delta x} = \lim_{\Delta x \to 0} \frac{f(x_0 + \Delta x, y_0) - f(x_0, y_0)}{\Delta x},$$

where $\Delta_x z = f(x_0 + \Delta x, y_0) - f(x_0, y_0)$ is the partial increment of the function with respect to x (in the general case, $\partial z/\partial x$ must not be regarded as a fraction; $\partial/\partial x$ is the symbol of an operation).

The partial derivative with respect to y is defined in a similar manner:

$$f'_y(x_0, y_0) = \lim_{\Delta y \to 0} \frac{\Delta_y z}{\Delta y} = \lim_{\Delta y \to 0} \frac{f(x_0, y_0 + \Delta y) - f(x_0, y_0)}{\Delta y}$$

where $\Delta_y z$ is the partial increment of the function with respect to y. Other notations include $\partial f(x_0, y_0)/\partial y$, $\partial f/\partial y$, z'_y, $\partial z/\partial y$, and $fy(x_0, y_0)$. Partial derivatives are calculated according to the rules of differentiation of functions of a single variable (in computing z'_x one assumes $y = \text{const}$ while if z'_y is calculated, one assumes $x = \text{const}$).

The partial differentials of $z = f(x, y)$ at (x_0, y_0) are, respectively,

$$d_x z = f_x^{'}(x_0, y_0)\, dx; \quad d_y z = f_y^{'}(x_0, y_0)\, dy,$$

where, as in the case of a single variable, $dx = \Delta x, dy = \Delta y$ denote the increments of the independent variables.

The first partial derivatives $\partial z / \partial x = f_x'(x, y)$ and $\partial z / \partial y = f_y'(x, y)$, or the partial derivatives of the first order, are functions of x and y, and may in their turn have partial derivatives with respect to x and y. These are named, with respect to the function $z = f(x, y)$, the partial derivatives of the second order, or second partial derivatives. It is assumed that

$$\frac{\partial}{\partial x}\left(\frac{\partial z}{\partial x}\right) = \frac{\partial^2 z}{\partial x^2}, \qquad \frac{\partial}{\partial y}\left(\frac{\partial z}{\partial y}\right) = \frac{\partial^2 z}{\partial x \partial y},$$

$$\frac{\partial}{\partial x}\left(\frac{\partial z}{\partial y}\right) = \frac{\partial^2 z}{\partial y \partial x}, \qquad \frac{\partial}{\partial y}\left(\frac{\partial z}{\partial y}\right) = \frac{\partial^2 z}{\partial y^2},$$

The following notations are also used instead of $\partial^2 z / \partial x^2$:

$$z_{xx}^{''},\ z_{x^2}^{''},\ \frac{\partial^2 f(x, y)}{\partial x^2},\ \frac{\partial^2 f}{\partial x^2}\ \ f_{xx}^{''}(x, y),\quad f_{x^2}^{''}(x, y),\quad f_{xx}(x, y);$$

and instead of $\partial^2 z / \partial x \partial y$:

$$z_{xy}^{''},\ \frac{\partial^2 f(x, y)}{\partial x \partial y},\ \frac{\partial^2 f}{\partial x \partial y},\qquad f_{xy}^{''}(x, y),\quad f_{xy}(x, y),$$

etc. One can introduce in the same manner partial derivatives of the third and higher orders, together with the respective notations: ,$\partial^n z / \partial x^n$ means that the function z is to be differentiated n times with respect to x; $\partial^n z / \partial x^p \partial y^q$ where $n = p + q$ means that the function z is differentiated p times with respect to x and q times with respect to y. The partial derivatives of second and higher orders obtained by differentiation with respect to different variables are known as mixed partial derivatives.

To each partial derivative corresponds some partial differential, obtained by its multiplication by the differentials of the independent variables taken to the powers equal to the number of differentiations with respect to the respective variable. In this way one obtains the n-th partial differentials, or the partial differentials of order n:

$$\frac{\partial^n z}{\partial x^n} dx^n, \qquad \frac{\partial^n z}{\partial x^p \partial y^q} dx^p\, d\, dy^q.$$

The following important theorem on derivatives is valid: If, in a certain neighbourhood of a point (x_0, y_0), a function $z = f(x, y)$ has mixed partial derivatives $f''_{xy}(x, y)$ and $f''_{yx}(x, y)$, and if these derivatives are continuous at the point (x_0, y_0), then they coincide at this point.

A function $z = f(x, y)$ is called differentiable at a point (x_0, y_0) with respect to both variables x and y if it is defined in some neighbourhood of this point, and if its total increment

$$\Delta z = f(x_0 + \Delta x, y_0 + \Delta y) - f(x_0, y_0)$$

may be represented in the form

$$\Delta z = A\,\Delta x + B\,\Delta y = \omega$$

where A and B are certain numbers and $\omega / \rho \to 0$ for $\rho = \sqrt{(\Delta x)^2 + (\Delta y)^2} \to 0$ (provided that the point $(x_0 + \Delta x, y_0 + \Delta y)$ lies in this neighbourhood). In this context, the expression

$$dz = df(x_0, y_0) = A\Delta x + B\Delta y$$

is called the total differential (of the first order) of f at (x_0, y_0) s; this is the principal linear part of increment. A function which is differentiable at a point is continuous at that point (the converse proposition is not always true!). Moreover, differentiability entails the existence of finite partial derivatives

$$f'_x(x_0, y_0) = \lim_{\Delta x \to 0} \frac{\Delta_x z}{\Delta x} = A, \qquad f'_y(x_0, y_0) = \lim_{\Delta y \to 0} \frac{\Delta_y z}{\Delta y} = B.$$

Thus, for a function which is differentiable at ,

$$dz = d\,f(x_0, y_0) = f'_x(x_0, y_0)\,\Delta x + f'_y(x_0, y_0)\,\Delta y$$

or

$$dz = d\,f(x_0, y_0) = f'_x(x_0, y_0)\,dx + f'_y(x_0, y_0)\,dy,$$

if, as in the case of a single variable, one puts, for the independent variables $dx = \Delta x, dy = \Delta y$.

The existence of finite partial derivatives does not, in the general case, entail differentiability (unlike in the case of functions in a single variable). The following is a sufficient criterion of the differentiability of a function in two variables: If, in a certain neighbourhood of a point (x_0, y_0), a function f has finite partial derivatives f'_x and f'_y which are continuous at (x_0, y_0), then f is differentiable at this point. Geometrically, the total differential $df(x_0, y_0)$ is the increment of the applicate of the tangent plane to the surface $z = f(x, y)$ at the point (x_0, y_0, z_0), where $z_0 = f(x_0, y_0)$ (cf. Figure).

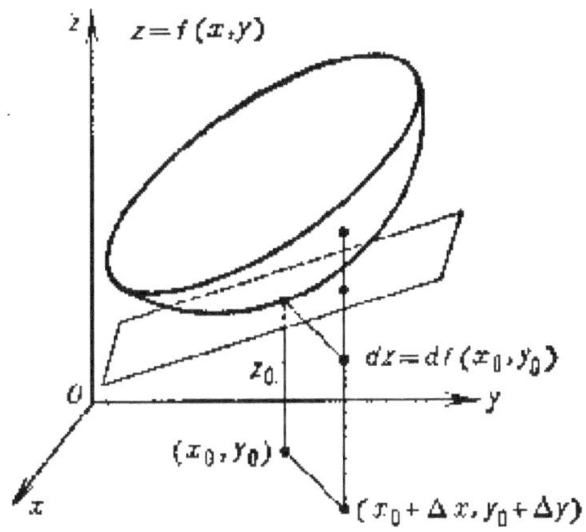

Figure

Total differentials of higher orders are, as in the case of functions of one variable, introduced by induction, by the equation

$$d^n z = d\,(d^{n-1}\, z),$$

on the assumption that the differential $d^{n-1}\, z$ is defined in some neighbourhood of the point under consideration, and that equal increments of the arguments $dx\,, dy$ are taken at all steps. Repeated differentials are defined in a similar manner.

Derivatives and Differentials of Composite Functions

Let $w = f\,(u_1\,,\dots\dots, u_m)$ be a function in m variables which is differentiable at each point of an open domain D of the m-dimensional Euclidean space R^m, and let m functions $u_1 = \phi_1\,(x_1\,,\dots., x_n)$ $,\dots, u_m = \phi_m\,(x_1,\dots, x_n)$ in n variables be defined in an open domain G of the n-dimensional Euclidean space R^n. Finally, let the point $(u_1\,,\dots\dots, u_m)$, corresponding to a point $(x_1\,,\dots., x_n) \in G$, be contained in D. The following theorems then hold:

A) If the functions $\phi_1\,,\dots.,\ \phi_m$ have finite partial derivatives with respect to , $x_1,\dots, x_n$ the composite function $w = f\,(u_1\,,\dots\dots, u_m)$ in $x_1,\dots, x_n$ also has finite partial derivatives with respect to $x_1,\dots, x_n$, and

$$\frac{\partial w}{\partial x_1} = \frac{\partial f}{\partial u_1}\frac{\partial u_1}{\partial x_1} + \dots\dots + \frac{\partial f}{\partial u_n}\frac{\partial u_n}{\partial x_1}$$

$$\frac{\partial w}{\partial x_n} = \frac{\partial f}{\partial u_1}\frac{\partial u_1}{\partial x_n} + \dots + \frac{\partial f}{\partial u_n}\frac{\partial u_n}{\partial x_n}$$

B) If the functions $\phi_1\ ,\dots, \phi_m$ are differentiable with respect to all variables at a point $(x_1,\dots, x_n) \in G$, then the composite function $w = f\,(u_1\,,\dots\dots, u_m)$ is also differentiable at that point, and

$$dw = \frac{\partial f}{\partial u_1}du_1 + \dots + \frac{\partial f}{\partial u_n}du_n\ ,$$

where du_1 ,..., du_m are the differentials of the functions u_1 ,..., u_m. Thus, the property of invariance of the first differential also applies to functions in several variables. It does not usually apply to differentials of the second or higher orders.

Differential calculus is also employed in the study of the properties of functions in several variables: finding extrema, the study of functions defined by one or more implicit equations, the theory of surfaces, etc. One of the principal tools for such purposes is the Taylor formula.

The concepts of derivative and differential and their simplest properties, connected with arithmetical operations over functions and superposition of functions, including the property of invariance of the first differential, are extended, practically unchanged, to complex-valued functions in one or more variables, to real-valued and complex-valued vector functions in one or several real variables, and to complex-valued functions and vector functions in one or several complex variables. In functional analysis the ideas of the derivative and the differential are extended to functions of the points in an abstract space.

Differentiable Function

A differentiable function is a function that can be approximated locally by a linear function.

Differentiability and Continuity

First we discuss the relation between differentiability and continuity.

Theorem: If $f:\ (a,\ b) \to R$ is differentiable at $c \in (a,\ b)$, then f is continuous at c.

Proof. If f is differentiable at c, then

$$\begin{aligned} \lim_{h\to 0} = f(c+h) - f(c) &= \lim_{h\to 0}\left[\frac{f(c+h)-f(c)}{h}\cdot h\right] \\ &= \lim_{h\to 0}\left[\frac{f(c+h)-f(c)}{h}\right].\lim_{h\to 0} h \\ &= f'(c)\cdot 0 \\ &= 0, \end{aligned}$$

which implies that f is continuous at c.

Definition. A function $f:\ (a,\ b) \to R$ is continuously differentiable on $(a,\ b)$, written $f \in C^1(a,\ b)$, if it is differentiable on $(a,\ b)$ and $f':\ (a,\ b) \to R$ is continuous.

For example, the function $f(x) = x^2$ with derivative $f'(x) = 2x$ is continuously differentiable on any interval $(a,\ b)$.

Differentiability Classes

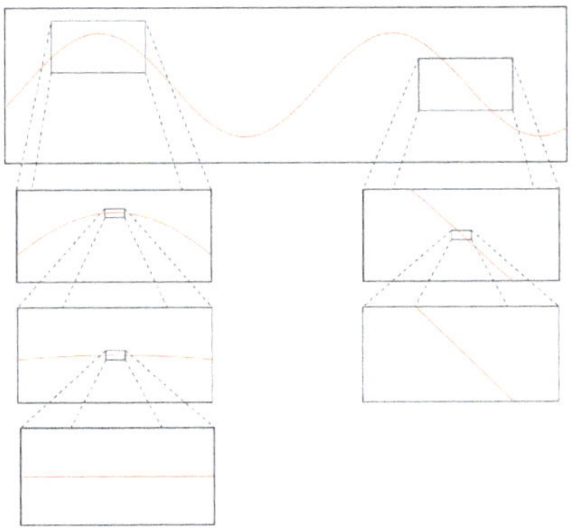

Differentiable functions can be locally approximated by linear functions.

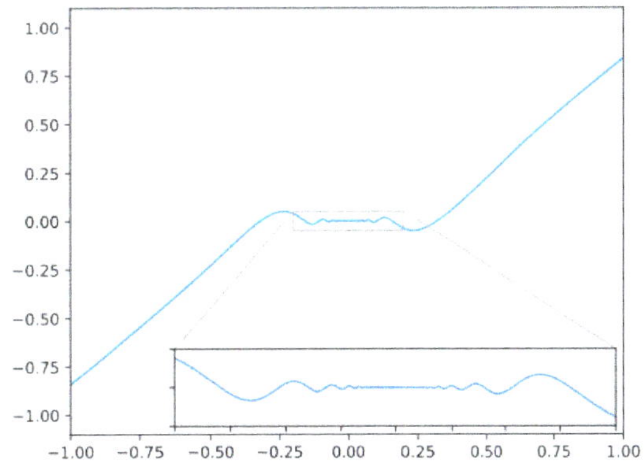

The function $f:\mathbb{R}\to\mathbb{R}$ with $f(x)=x^2\sin\left(\frac{1}{x}\right)$ for $x\neq 0$ and $f(0)=0$ is differentiable. However, this function is not continuously differentiable.

A function *f* is said to be continuously differentiable if the derivative *f′*(*x*) exists and is itself a continuous function. Though the derivative of a differentiable function never has a jump discontinuity, it is possible for the derivative to have an essential discontinuity. For example, the function

$$f(x)=\begin{cases} x^2\sin(1/x) & \text{if } x\neq 0 \\ 0 & \text{if } x=0 \end{cases}$$

is differentiable at 0, since

$$f'(0)=\lim_{\epsilon\to 0}\left(\frac{\epsilon^2\sin(1/\epsilon)-0}{\epsilon}\right)=0,$$

exists. However, for $x\neq 0$,

$$f'(x)=2x\sin(1/x)-\cos(1/x)$$

which has no limit as $x\to 0$. Nevertheless, Darboux's theorem implies that the derivative of any function satisfies the conclusion of the intermediate value theorem.

Sometimes continuously differentiable functions are said to be of class C^1. A function is of class C^2 if the first and second derivative of the function both exist and are continuous. More generally, a function is said to be of class C^k if the first *k* derivatives $f'(x), f''(x), \ldots, f^{(k)}(x)$ all exist and are continuous. If derivatives $f^{(n)}$ exist for all positive integers n, the function is smooth or equivalently, of class C^∞.

Differentiability in Higher Dimensions

If all the partial derivatives of a function exist and are continuous in a neighborhood of a point, then the function is differentiable at that point, and it is of class C^1.

Formally, a function of several real variables $f: R^m \to R^n$ is said to be differentiable at a point x_0 if there exists a linear map $J: R^m \to R^n$ such that

$$\lim_{\mathbf{h}\to\mathbf{0}} \frac{\|\mathbf{f}(\mathbf{x}_0+\mathbf{h})-\mathbf{f}(\mathbf{x}_0)-\mathbf{J}(\mathbf{h})\|_{\mathbf{R}^n}}{\|\mathbf{h}\|_{\mathbf{R}^m}} = 0.$$

If a function is differentiable at x_0, then all of the partial derivatives exist at x_0, in which case the linear map **J** is given by the Jacobian matrix. A similar formulation of the higher-dimensional derivative is provided by the fundamental increment lemma found in single-variable calculus.

Note that existence of the partial derivatives (or even all of the directional derivatives) does not in general guarantee that a function is differentiable at a point. For example, the function $f: R^2 \to R$ defined by

$$f(x,y) = \begin{cases} x & \text{if } y \neq x^2 \\ 0 & \text{if } y = x^2 \end{cases}$$

is not differentiable at (0, 0), but all of the partial derivatives and directional derivatives exist at this point. For a continuous example, the function

$$f(x,y) = \begin{cases} y^3/(x^2+y^2) & \text{if } (x,y) \neq (0,0) \\ 0 & \text{if } (x,y) = (0,0) \end{cases}$$

is not differentiable at (0, 0), but again all of the partial derivatives and directional derivatives exist.

Differentiability in Complex Analysis

In complex analysis, any function that is complex-differentiable in a neighborhood of a point is called holomorphic. Such a function is necessarily infinitely differentiable, and in fact analytic.

Differentiable Functions on Manifolds

If M is a differentiable manifold, a real or complex-valued function f on M is said to be differentiable at a point p if it is differentiable with respect to some (or any) coordinate chart defined around p. More generally, if M and N are differentiable manifolds, a function $f: M \to N$ is said to be differentiable at a point p if it is differentiable with respect to some (or any) coordinate charts defined around p and $f(p)$.

Differential of a Function

Definition of the Differential of a Function

Consider a function $y = f(x)$ which is continuous in the interval $[a,b]$. Suppose that at some point $x_0 \in [a,b]$ the independent variable is incremented by Δx. The increment of the function Δy corresponding to the change of the independent variable Δx is given by

$$\Delta y = \Delta f(x_0) = f(x_0 + \Delta x) - f(x_0).$$

For any differentiable function, the increment Δy can be represented as a sum of two terms:

$$\Delta y = A\Delta x + \text{o}(\Delta x),$$

where the first term (called the *principal part* of the increment) is linearly dependent on the increment Δx and the second term has a higher order of smallness with respect to Δx. The expression A Δx is called the *differential of function* and is denoted by dy or $df(x_0)$.

Consider the idea of partition of the increment of of the function Δy into two parts in the following simple example. Given a square with side $x_0 = 1\,m$ (Figure). Its area is obviously equal to

$$S_0 = x_0^2 = 1\,m^2.$$

If the side of the square is increased by $\Delta x = 1\,cm$,the exact value of the area of the square will be equal to

$$S = x^2 = (x_0 + \Delta x)^2 = 1,01^2 = 1,0201\,m^2,$$

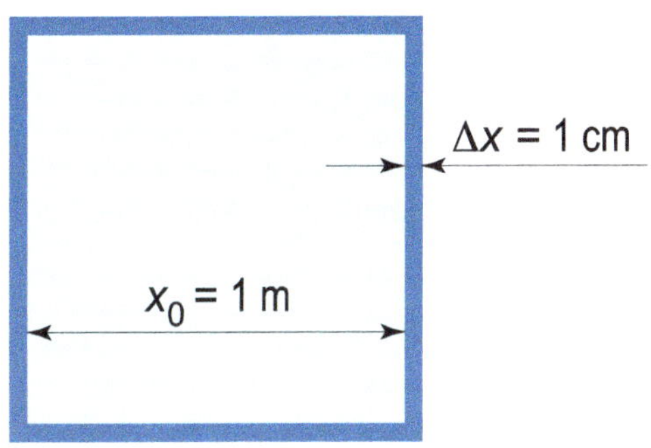

Figure

i.e. the increment of the area ΔS is

$$\Delta S = S - S_0 = 1,0201 - 1 = 0,0201\,m^2 = 201\,cm^2.$$

We now represent this increment ΔS as follows:

$$\Delta S = S - S_0 = (x_0 + \Delta x)^2 - x_0^2 = x_0^2 + 2x_0\Delta x + (\Delta x)^2 - x_0^2$$
$$= 2x_0\Delta x + (\Delta x)^2 = A\Delta x + o(\Delta x) = dy + o(\Delta x).$$

Thus, the increment ΔS consists of the principal part (the differential of the function), which is proportional to Δx and is equal to

$$dy = A\Delta x = 2x_0\Delta x = 2 \cdot 1 \cdot 0,01 = 0,02\,m^2 = 200\,cm^2,$$

and the term of a higher order of smallness, which in turn is equal to

$$o(\Delta x) = (\Delta x)^2 = 0,01^2 = 0,0001\, m^2 = 1\, cm^2.$$

In sum, both these terms comprise the full increment of the square area equal to $200 + 1 = 201\, cm^2$.

In this example the coefficient A is equal to the value of the derivative of S at the point x_0:

$$A = 2x_0.$$

It turns out that for any differentiable function, the following *theorem* is valid:

The coefficient A in the principal part of the increment of a function at a point x_0 is equal to the value of the derivative at this point, i.e. the increment Δy is given by

$$\Delta y = A\Delta x + o(\Delta x) = f'(x_0)\Delta x + o(\Delta x).$$

Dividing both sides of the equation by $\Delta x \neq 0$ gives

$$\frac{\Delta y}{\Delta x} = A + \frac{o(\Delta x)}{\Delta x} = f'(x_0) + \frac{o(\Delta x)}{\Delta x}$$

In the limit as $\Delta x \to 0$,we obtain the value of the derivative at the point x_0:

$$y'(x0) = \lim_{\Delta x \to 0} \frac{\Delta y}{\Delta x} = A = f'(x0).$$

Here we took into account that for a small quantity $o(\Delta x)$ of higher order of smallness than Δx, the limit is equal to

$$\lim_{\Delta x \to 0} \frac{o(\Delta x)}{\Delta x} = 0$$

Assuming that the *differential of the independent variable dx* is equal to its increment Δx:

$$dx = \Delta x$$

we obtain from the relation

$$dy = A\Delta x = y'dx$$

that

$$y' = \frac{dy}{dx},$$

i.e. the derivative of a function can be represented as the ratio of two differentials.

Geometric Meaning of the Differential of a Function

Figure below schematically shows splitting of the increment Δy into the principal part $A\Delta x$ (the differential of function) and the term of a higher order of smallness $o(\Delta x)$.

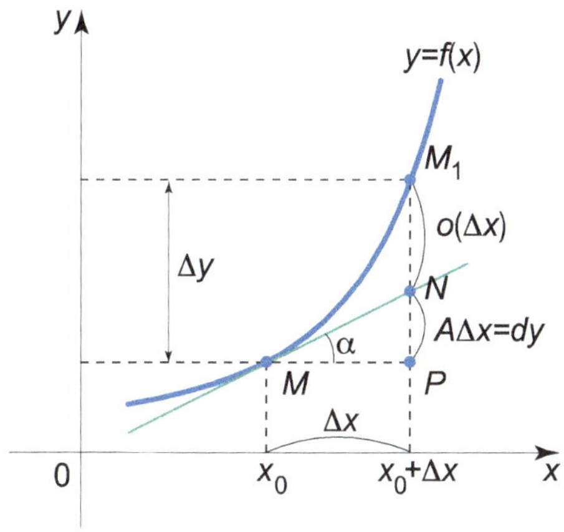

Figure

The tangent MN drawn to the curve of the function $y = f(x)$ at the point M, as it is known, has the slope angle α, the tangent of which is equal to the derivative:

$$tan\,\alpha = f'(x_0).$$

When the independent variable changes by Δx, the tangent increments by $A\Delta x$. This linear increment formed by the tangent is just the differential of the function. The remaining part of the full increment Δy (the segment NM_1) corresponds to the "nonlinear" additive of a higher order of smallness with respect to Δx.

Properties of the Differential

Let u and v be functions of the variable x. The differential has the following properties:

1. A constant can be taken out of the differential sign:
   $d(Cu) = Cdu$, where C is a constant number.

2. The differential of the sum (difference) of two functions is equal to the sum (difference) of their differentials:
   $d(u \pm v) = du \pm dv$.

3. The differential of a constant is zero:
   $d(C) = 0.$

4. The differential of the independent variable x is equal to its increment:
   $dx = \Delta x.$

5. The differential of a linear function is equal to its increment:
   $d(ax+b) = \Delta(ax+b) = a\Delta x.$

6. Differential of the product of two functions:

   $$d(uv) = du \cdot v + u \cdot dv.$$

7. Differential of the quotient of two functions:

   $$d\left(\frac{u}{v}\right) = \frac{du \cdot v - u \cdot dv}{v^2}.$$

8. The differential of a function is equal to the derivative of the function times the differential of the independent variable:

   $$dy = df(x) = f'(x)dx.$$

As can be seen, the differential of the function dy differs from the derivative only by the factor dx. For example,

$$d(x^n) = nx^{n-1}dx, d(\ln x) = \frac{dx}{x}, d(\sin x) = \cos x dx$$

and so on.

Form Invariance of the Differential

Consider a composition of two functions $y = f(u)$ and $u = g(x)$. Its derivative can be found by the chain rule:

$$y'_x = y'_u \cdot u'_x,$$

where the subindex denotes the variable of differentiation.

The differential of the "outer" function $y = f(u)$ can be written as

$$dy = y'_u du.$$

The differential of the "inner" function $u = g(x)$ can be represented in a similar manner:

$$du = u'_x\ dx.$$

If we substitute du in the last formula, we obtain

$$dy = y'_u du = y'_u\ u'_x dx.$$

Since $y'_x = y'_u\ u'_x$, then

$$dy = y'_x dx.$$

It is seen that in the case of a composite function, we obtain the expression for the differential in the same form as for a "simple" function. This property is called the *form invariance of the differential.*

Implicit Function

The Implicit function is a function in which the dependent variable has not been given "explicitly" in terms of the independent variable. Or it is a function in which the dependent variable is not expressed in terms of some independent variables.

It is denoted by:

$$R(x,y) = 0$$

Some examples of Implicit Functions are:

$$x^2 + y^2 - 1 = 0$$

$$y^4 + x^3 + 17 = 0$$

Although you can convert a Implicit function into Explicit function it is generally not done because after conversion.

In general, we are interested in studying relations in which one function of x and y is equal to another function of x and y. A function f of x and y takes each ordered pair (x, y) and associates it to some number f(x, y). A general way to write down the type of relations in which we are interested is:

$$f(x, y) = g(x, y)$$

The relation $x^2 + y^2 = 1$ which defines the circle of radius 1 centered at the origin is one such relation: in this case, $f(x, y) = x^2 + y^2$ and $g(x, y)$ is the constant function 1. Another such relation is $y - 1 x^2 + 2x$. This relation also defines a curve, a parabola. How do we see this? The natural thing to do is to solve for y:

$$y - 1 = x^2 + 2x$$

$$y = x^2 + 2x + 1$$

Thus we see that the curve defined by the relation $y - 1 = x^2 + 2x$ is just the graph of a quadratic function. The function $y = x^2 + 2x + 1$ that we found by solving for y is called the implicit function of the relation $y - 1 = x^2 + 2x$. In general, any function we get by taking the relation $f(x, y) = g(x, y)$ and solving for y is called an implicit function for that relation.

Examples

Inverse Functions

A common type of implicit function is an inverse function. If g is a function of x, then the inverse function of g, called g^{-1}, is the function giving a solution of the equation

$$y = g(x)$$

for x in terms of y. This solution is

$$x = g^{-1}(y).$$

Here g^{-1} is an implicit function, and more specifically an inverse function. While for some functions g, g^{-1} could be written out explicitly — for instance, if $g(x) = 2x$, then $g^{-1}(y) = \frac{1}{2}y$ — this is often not the case (as in the product log example below).

Intuitively, an inverse function is obtained from g by interchanging the roles of the dependent and independent variables. Stated another way, the inverse function gives the implicit solution for x of the equation

$$R(x, y) = y - g(x) = 0.$$

Example

The product log is an implicit function giving the solution for x of the equation $y - xe^x = 0$.

Algebraic Functions

An algebraic function is a function that satisfies a polynomial equation whose coefficients are themselves polynomials. For example, an algebraic function in one variable x gives a solution for y of an equation

$$a_n(x)y^n + a_{n-1}(x)y^{n-1} + \cdots + a_0(x) = 0$$

where the coefficients $a_i(x)$ are polynomial functions of x. This algebraic function can be written as the right side of the solution equation $y = f(x)$. Written like this, f is a multi-valued implicit function.

Algebraic functions play an important role in mathematical analysis and algebraic geometry. A simple example of an algebraic function is given by the left side of the unit circle equation:

$$x^2 + y^2 - 1 = 0.$$

Solving for y gives an explicit solution:

$$y = \pm\sqrt{1 - x^2}.$$

But even without specifying this explicit solution, it is possible to refer to the implicit solution of the unit circle equation as $y = f(x)$, where f is the multi-valued implicit function.

While explicit solutions can be found for equations that are quadratic, cubic, and quartic in y, the same is not in general true for quintic and higher degree equations, such as

$$y^5 + 2y^4 - 7y^3 + 3y^2 - 6y - x = 0.$$

Nevertheless, one can still refer to the implicit solution $y = f(x)$ involving the multi-valued implicit function f.

Caveats

Not every equation $R(x,y) = 0$ implies a graph of a single-valued function, the circle equation being one prominent example. Another example is an implicit function given $x - C(y) = 0$ where C is a cubic polynomial having a "hump" in its graph. Thus, for an implicit function to be a *true* (single-valued) function it might be necessary to use just part of the graph. An implicit function can sometimes be successfully defined as a true function only after "zooming in" on some part of the *x*-axis and "cutting away" some unwanted function branches. Then an equation expressing *y* as an implicit function of the other variable(s) can be written.

The defining equation $R(x,y) = 0$ can also have other pathologies. For example, the equation $x = 0$ does not imply a function $f(x)$ giving solutions for *y* at all; it is a vertical line. In order to avoid a problem like this, various constraints are frequently imposed on the allowable sorts of equations or on the domain. The implicit function theorem provides a uniform way of handling these sorts of pathologies.

Implicit differentiation is to treat y as a function of x. To emphasize this, let us rewrite the relation above, replacing *y* with $y(x)$:

$$\sin(y(x)) = x.$$

Now we differentiate each side of this equation, and set their derivatives equal to each other. Since we do not know the formula for $y(x)$, we just leave its derivative as $y'(x)$:

$$\cos(y(x)) \,.\, y'(x) = 1.$$

Finally, we solve for $y'(x)$ to get its formula:

$$y'(x) = \frac{1}{\cos(y(x)} = \frac{1}{\cos y}.$$

At the end we turn y (x) back into y to make the notation less cumbersome. What this tells us is that, even though we do not have formulae for the implicit functions of sin $\sin y = x$, we know that the derivative of thoseimplicit functions is given by $\frac{1}{\cos x}$. So, for example, we know that the Point $\left(\frac{1}{2}, \frac{\pi}{6}\right)$ is a point on the curveof the relation $\mathit{siny} = x$:

$$\sin y = \sin\left(\frac{\pi}{6}\right) = \frac{1}{2} = x.$$

Whatever the formula for the implicit function of $\sin y = x$ at $\left(\frac{1}{2}, \frac{\pi}{6}\right)$ is, we know that its derivative at that point is

$$y'(x) = \frac{1}{\cos\left(\frac{\pi}{6}\right)} = \frac{1}{\frac{\sqrt{3}}{2}} = \frac{2}{\sqrt{3}}$$

We need to analyze the relation $\sin y = x$ and its derivative a bit more, but first let us go back to our first example, the relation $x^2 + y^2 = 1$. In this case, we know how to find the implicit functions of this relation, but let us use implicit differentiation anyway. First, we rewrite this relation by replacing y with y (x) :

$$x^2 + (y(x))^2 = 1.$$

Next, we differentiate both sides and set the derivatives equal to each other:

$$2x + 2y(x) \cdot y'(x) = 0$$

Finally we solve for $y'(x)$.:

$$2x + 2y(x) \cdot y'(x) = 0$$

$$2y(x) \cdot y'(x) = -2x$$

$$y'(x) = \frac{-2x}{2y(x)}$$

$$y'(x) = -\frac{x}{y}.$$

Before we said that one point on this circle is $(0, 1)$. This formula for the derivative of the implicit function tells that the slope of the curve at that point should be 0, and this is precisely what we see when draw out the circle $x^2 + y^2 = 1$ and plot the point $(0, 1)$.. Another point on this circle is $(1, 0)$. We considered this a bad point, because it had more than one implicit function associated to it. When we try to apply our formula for $y'(x)$ to this point, we see that the formula is undefined at $(1, 0)$.. This makes sense: the tangent line to the circle $x^2 + y^2 = 1$ at $(1, 0)$. is a vertical line, so its slope is undefined. What happened here is that the formula for $y'(x)$ is a quotient, and at $(1, 0)$., the denominator is 0. This gives us a way to find the bad points of a relation, the points where the implicit function for the relation is not well-defined: a bad point is a point where the formula for $y'(x)$ is undefined. In this case, the bad points for the relation $x^2 + y^2 = 1$ are the points on the circle where $y = 0$, so that

$$x^2 + y^2 = 1$$

$$x^2 + 0^2 = 1$$

$$x^2 = 1$$

$$x = 1 \text{ or } x = -1.$$

So the bad points are $(1, 0)$. and $(-1, 0)$, which is exactly what we would expect looking at the circle or radius 1 centered at the origin.

Going back to the relation $\sin y = x$, let us try to find the bad points of this relation. The formula for the derivative that we got before is

$$y'(x) = \frac{1}{\cos y},$$

so the bad points will be where $\cos y = 0$. So the bad points have y-coordinates

$$\ldots, -\frac{3\pi}{2}, -\frac{\pi}{2}, \frac{\pi}{2}, \frac{3\pi}{2}, \ldots.$$

Since $x = \sin y$ by the formula for the relation, we get that the bad points of $\sin y = x$ are the points

$$\ldots, \left(\frac{3\pi}{2}, 1\right), \left(-\frac{\pi}{2}, -1\right), \left(\frac{\pi}{2}, 1\right), \left(\frac{3\pi}{2}, -1\right), \ldots.$$

Let us do one more example of implicit differentiation using the conic section

$$x^2 + 2xy + y^2 - y + x = 0.$$

Replacing y with $y(x)$, we get

$$x^2 + 2x \cdot y(x) + (y(x))^2 - y(x) + x = 0.$$

Differentiating both sides and setting the derivatives equal to each other, we have that

$$2x + 2y(x) + 2x \cdot y'(x) + 2y(x) \cdot y'(x) - y'(x) + 1 = 0.$$

Now we solve for $y'(x)$:

$$2x + 2y(x) + 2x \cdot y'(x) + 2y(x) \cdot y'(x) - y'(x) + 1 = 0.$$

$$(2x + 2y(x) + 1) + (2x + 2y(x) - 1) \cdot y'(x) = 0$$

$$(2x + 2y(x) - 1) \cdot y'(x) = -(2x + 2y(x) + 1)$$

$$y'(x) = \frac{-(2x + 2y(x) + 1)}{(2x + 2y(x) - 1)}$$

$$y'(x) = \frac{2x + 2y + 1}{2x + 2y - 1}$$

So, for example, the derivative of the implicit function of this conic section at the point $(-3;6)$ is

$$y'(x) = \frac{2x + 2y + 1}{2x + 2y - 1} = -\frac{2\cdot(-3) + 2\cdot 6 + 1}{2\cdot(-3) + 2\cdot 6 - 1} = -\frac{-6 + 12 + 1}{-6 + 12 - 1} = \frac{7}{5}.$$

Try differentiating the formula for the implicit function that we got before to verify this value for the derivative at $(-3, 6)$. The bad points of this conic section are when $2x + 2y - 1 = 0$. If we were being industrious, we could solve for the bad points explicitly, but let us leave that alone for now.

Implicit Differentiation

In calculus, a method called implicit differentiation makes use of the chain rule to differentiate implicitly defined functions.

To differentiate an implicit function $y(x)$, defined by an equation $R(x, y) = 0$, it is not generally possible to solve it explicitly for y and then differentiate. Instead, one can totally differentiate $R(x, y) = 0$ with respect to x and y and then solve the resulting linear equation for dy/dx to explicitly get the derivative in terms of x and y. Even when it is possible to explicitly solve the original equation, the formula resulting from total differentiation is, in general, much simpler and easier to use.

Examples

1. Consider for example

$$y + x + 5 = 0.$$

This equation is easy to solve for y, giving

$$y = -x - 5,$$

where the right side is the explicit form of the function $y(x)$. Differentiation then gives $dy/dx = -1$. Alternatively, one can totally differentiate the original equation:

$$\frac{dy}{dx} + \frac{dx}{dx} + \frac{d}{dx}(5) = 0;$$

$$\frac{dy}{dx} + 1 = 0.$$

Solving for dy/dx gives

$$\frac{dy}{dx} = -1,$$

the same answer as obtained previously.

2. An example of an implicit function for which implicit differentiation is easier than using explicit differentiation is the function $y(x)$ defined by the equation

$$x^4 + 2y^2 = 8.$$

To differentiate this explicitly with respect to *x*, one has first to get

$$y(x) = \pm\sqrt{\frac{8 - x^4}{2}},$$

and then differentiate this function. This creates two derivatives: one for ≥ 0 and another for $y < 0$.

It is substantially easier to implicitly differentiate the original equation:

$$4x^3 + 4y\frac{dy}{dx} = 0,$$

giving

$$\frac{dy}{dx} = \frac{-4x^3}{4y} = \frac{-x^3}{y}.$$

3. Often, it is difficult or impossible to solve explicitly for *y*, and implicit differentiation is the only feasible method of differentiation. An example is the equation

$$y^5 - y = x.$$

It is impossible to algebraically express *y* explicitly as a function of *x*, and therefore one cannot find dy/dx by explicit differentiation. Using the implicit method, dy/dx can be obtained by differentiating the equation to obtain

$$5y^4\frac{dy}{dx} - \frac{dy}{dx} = \frac{dx}{dx}$$

where $dx/dx = 1$. Factoring out dy/dx shows that

$$(5y^4 - 1)\frac{dy}{dx} = 1$$

which yields the result

$$\frac{dy}{dx} = \frac{1}{5y^4 - 1},$$

which is defined for $y \neq \pm\frac{1}{\sqrt[4]{5}}$ and $y \neq \pm i\frac{1}{\sqrt[4]{5}}$.

General Formula for Derivative of Implicit Function

If $R(x, y) = 0$, the derivative of the implicit function $y(x)$ is given by

$$\frac{dy}{dx} = -\frac{\partial R / \partial x}{\partial R / \partial y} = -\frac{R_x}{R_y},$$

where R_x and R_y indicate the partial derivatives of R with respect to x and y.

The above formula comes from using the generalized chain rule to obtain the total derivative—with respect to x—of both sides of $R(x, y) = 0$,

$$\frac{\partial R}{\partial x}\frac{dx}{dx} + \frac{\partial R}{\partial y}\frac{dy}{dx} = 0,$$

hence

$$\frac{\partial R}{\partial x} + \frac{\partial R}{\partial y}\frac{dy}{dx} = 0,$$

which, when solved for dy / dx, gives the expression above.

Theorem. Suppose F(x, y) is continuously differentiable in a neighborhood of a point (a, b) $\in \mathbb{R}^n \times \mathbb{R}$ and F(a, b) = 0. Suppose that F_y(a, b) $\neq$ 0. Then there is $\delta > 0$ and $\in > 0$ and a box B = $\{(x, y) : \| x - a \| < \delta, |y - b| < \in\}$ so that

1. For each x such that $\| x - a \| < \delta$ there is a unique y such that $|y - b| < \in$ for which $F(x, y) = 0$.

This correspondence defines a function $f(x)$ on $\{\| x - a \| < \delta\}$ such that

$$F(x, y) = 0 \Leftrightarrow y = f(x) \text{ for } (x, y) \in B$$

2. f is continuous.

3. f is continuously *differentiable* and

$$Df(x) = -\frac{D_x F(x, f(x))}{F_y(x, f(x))},$$

where $Df \quad [f_x \ldots\ldots, f_{xn}]$ and $D_y F = [F_{x1} \ldots\ldots, F_{xn}]$

Proof.

1. Choose $\delta_1 > 0$ and $\varepsilon_1 > 0$ so that $F_y(x, y) > 0$ for $\| x - a \| < \delta_1$; $|y - b| < \varepsilon_1$. Since $F(a, b) = 0$ and $F(a, y)$ is strictly increasing in y, F(a, b $+\varepsilon_1$=2) > 0 and F(a, b $-\varepsilon_1$=2) > 0. Let $\varepsilon = \varepsilon_1/2$ and choose $\delta < \delta_1$ so that F(x, b +) > 0 and F(x, b ..) < 0 if F(x, b ..) < δ. These dimensions define B. For fixed x with $\| x - a \| < \delta$, since F(x, b $-\varepsilon$) < 0, F(x, b + ε) > 0, and $F(x, y)$ is strictly increasing in y, the intermediate value theorem implies that there is a unique y with $|y - b| < \varepsilon_1$ such that $F(x, y) = 0$. The uniquely determined y defines a function $f(x)$. This proves the first statement.

2. We prove that f is continuous at a. Let $e > 0$ be given. Assume that $e < \in$ Then by the proof of the first statement, there is a $d > 0$ (we may choose $d < \delta$) so that the uniquely defined $f(x)$ in $\{\| x - a \| < d\}$ satisfies $|f(x) - b| < d$. This proves continuity at a. We can repeat this argument at any point $(a_1, f(a_1)) \in B$, proving that f is continuous on $\{\| x - a \| < \in\}$.

3. By differentiability

$$0 = \text{F}(\text{x, f(x)}) = \text{F}(\text{a, b}) + \sum \text{P}\ (\text{x}, f(\text{x}))(\text{x - a}\ ) + \text{Q}(\text{x}, f(\text{x}))(f(\text{x}) - f(\text{a}))$$

$$= \sum_j \text{P}_j(\text{x}, f(\text{x}))(\text{x - a}_\text{j}) + \text{Q}(\text{x}, f(\text{x}))(f(\text{x}) - f(\text{a})),$$

where $\text{P}\ (\text{x}, f(\text{x}))$, $\text{Q}(\text{x}, f(\text{x}))$ are continuous at a . Rewrite this as

$$\text{Q}(\text{x}, f(\text{x}))(f(\text{x}) - f(\text{a})) = -\sum_j \text{P}_j(\text{x}, f(\text{x}))(\text{x -a}_\text{j}\ ).$$

Since $\text{Q}(\text{x}, f(\text{x}))$ is continuous at a and $Q(a, (f(a)) = f_y(a,b) > 0, Q(x, f(x)) > 0$ for x near a and we can divide by it to get

$$f(x) = f(a) + -\sum_j \frac{\text{P}_j(\text{x}, f(\text{x}))}{\text{Q}(\text{x}, f(\text{x}))}(\text{x - a}_\text{j}).$$

Each term $\dfrac{\text{P}_j(\text{x}, f(\text{x}))}{\text{Q}(\text{x}, f(\text{x}))}$ is continuous at a so f is differentiable at a . Moreover

$$f_{xj}\ (a,b) = \frac{F_j\ (a,b)}{F_y(a,b)}.$$

You might like this bad notation:

$$\frac{\partial y}{\partial x_j} = \frac{\partial F_{xj}}{\partial F_y}.$$

Theorem (Special Implicit Function Theorem) Suppose that $\text{F} : \text{R}^{n+1} \to \text{R}$ has continuous partial derivatives. Denoting points in R^{n+1} by (x, z), where $\in \text{R}^n$ and $\text{z} \in \text{R}$, assume that (x_0, z_0) satisfies

$$\text{F}(\text{x}_0, \text{z}_0) = 0 \text{ and } \frac{\partial F}{\partial_z}(\text{x}_0, \text{z}_0) \neq 0.$$

Then there is an open ball $\text{U} \subset \text{R}^n$ containing x_0 and an interval $V \subset R$ containing z_0 such that there is a unique function $\text{z} = \text{g}(\text{x})$ defined for $\text{x} \in \text{U}$ and $\text{z} \in \text{V}$ that satisfies $\text{F}(\text{x, g(bx)}) = 0$.

Moreover, if $x \in U$ and $z \in V$ satisfy $F(x, z) = 0,$, then $z = g(x)$. Finally, $z = g(x)$ is continuously differentiable, with the derivative given by

$$Dg(x) = \frac{1}{\frac{\partial F}{\partial_z}(x, z)} D_x F(x, z) \Bigg|_{z=g(x)}$$

where $D_x F = [\partial F/\partial x_1 \ldots \partial F/\partial x_n]$, in other words

$$\frac{\partial g}{\partial x_i} = -\frac{\partial F/\partial x_1}{\partial F/\partial z}, \quad i = 1, \ldots, n$$

You will find a proof of this an a standard Advanced Calculus book. We will not prove it here.

The general theorem gives us a system of equations in several variables that we must solve. What are the criteria for deciding when we can solve for some of the variables in terms of the others, or when such an implicit function can be found?

Assume that we are given a system of n equations

$$\begin{aligned} f_1(x_1, \ldots, x_n, y_1, \ldots, y_m) &= 0 \\ f_2(x_1, \ldots, x_n, y_1, \ldots, y_m) &= 0 \\ \vdots \qquad \vdots \qquad \vdots \quad & \vdots \\ f_n(x_1, \ldots, x_n, y_1, \ldots, y_m) &= 0 \end{aligned}$$

in $n+m$ variables $x_1, \ldots, x_n, y_1, \ldots, y_m$. We want sufficient conditions so that we can solve for $x_1, \ldots, x_n$ in terms of $y_1, \ldots, y_m$.

Theorem: (Implicit Function Theorem) If $f_1, \ldots, f_n$ are differentiable functions on a neighborhood of the point $(x_0, y_0) = (x_1^0, \ldots, x_n^0, y_1^0, \ldots, y_m^0)$ in $R^n \times R^m$, if $f_1(x_0, y_0) = f_2(x_0, y_0) = \ldots = f_n(x_0, y_0) = 0$ and if the $n \times n$ matrix

$$\begin{pmatrix} \frac{\partial f_1}{\partial x_1} & \frac{\partial f_1}{\partial_{x_2}} & \cdots & \frac{\partial f_1}{\partial_{x_n}} \\ \frac{\partial f_2}{\partial x_1} & \frac{\partial f_2}{\partial x_2} & \cdots & \frac{\partial f_2}{\partial_{x_n}} \\ \vdots & \vdots & \ddots & \vdots \\ \frac{\partial f_n}{\partial x_1} & \frac{\partial f_n}{\partial x_2} & \cdots & \frac{\partial f_n}{\partial x_n} \end{pmatrix}$$

is nonsingular at (x_0, y_0), then there is a neighborhood U of the point $y_0 = (y_1^0, \ldots, y_m^0)$ in R^m, there is a neighborhood V of the point $x0 = (x_1^0, \ldots, x_n^0$ in R^n, and there is a unique mapping

$\varphi\, U \to V$ such that $\varphi\,(y_0) = x_0$ and $f_1(\varphi(y), y) = \ldots = f_n(\varphi(y), y) = 0$ for all y in U. Furthermore, 'φ is differentiable.

This says that if we write

$$\varphi(y) = (g_1(y), \ldots, g_n(y))$$

where $g_1, \ldots, g_n$ are differentiable functions on U, then

$$\begin{aligned} x_1 &= g_1(y_1, \ldots, y_m) \\ &\vdots \\ x_n &= g_n(y_1, \ldots, y_m) \end{aligned}$$

is the unique solution to the above system of equations near y_0.

If we restrict to a special case, namely n = 3 and m = 1, the Implicit Function Theorem gives us the following corollary.

Corollary 1 Let $f : R^3 \to R$ be a given function having continuous partial derivatives. Suppose that (x_0, y_0, z_0) is a point satisfying

$$f(x_0, y_0, z_0) = 0 \text{ and } \frac{\partial f}{\partial z}(x_0, y_0, z_0) \neq 0$$

Then there is an open disk $U \subset R^2$ containing (x_0, y_0) and an open interval V in R containing z_0 such that there is a unique function $z = g(x, y)$ defined for $(x, y) \in U$ and $z \in V$ that satisfies

$$f(x, y, g(x, y)) = 0.$$

Moreover, $z = g(x, y)$ is differentiable with the derivative given by

$$\begin{pmatrix} \dfrac{\partial g}{\partial x} = -\dfrac{\partial f}{\partial x} \Big/ \dfrac{\partial f}{\partial z} \\ \dfrac{\partial g}{\partial y} = -\dfrac{\partial f}{\partial y} \Big/ \dfrac{\partial f}{\partial z} \end{pmatrix}$$

We shall find this useful when it comes to parametrizing surfaces and doing "patchwork"

A special case of the Implicit Function Theorem is the Inverse Function Theorem. In this case we are trying to solve the n equations

$$
\begin{aligned}
f_1(x_1, \ldots, x_n) &= y_1 \\
f_2(x_1, \ldots, x_n) &= y_2 \\
\vdots \quad & \quad \vdots \\
f_n(x_1, \ldots, x_n) &= y_n
\end{aligned}
$$

for $x_1, \ldots, x_n$ as functions of $y_1, \ldots, y_n$; that is, we are trying to invert the equations in this system. The question of when we can solve these equations is answered by the general implicit function theorem applied to the functions $y_i - fi(x1, \ldots, x_n)$. The condition for solvability in a neighborhood of a point xo is that the determinant of the matrix $Df(x_0)$ is nonzero. This determinant is called the Jacobian determinant of f and is

$$
\frac{\partial(f_1, \ldots, f_n)}{\partial(x_1, \ldots, x_n)} = J(f)\begin{vmatrix} \frac{\partial f_1}{\partial x_1} \cdots \frac{\partial f_1}{\partial x_n} \\ \vdots \quad\quad \vdots \\ \frac{\partial f_n}{\partial x_1} \cdots \frac{\partial f_n}{\partial x_n} \end{vmatrix}
$$

Theorem: (Inverse Function Theorem) Let $U \subset R^n$ be open and let $f_i : U \to R$ have continuous partial derivatives. Consider the equations near a solution x_0, y_0. If the Jacobian determinant, $J(f)(x_0) \neq 0$, then it can be solved uniquely as $x = g(y$ for x near x_0 and y near y_0. Moreover, the function g has continuous partial derivatives.

In Algebraic Geometry

Consider a relation of the form $R(x_1,\ldots,x_n) = 0$, where *R* is a multivariable polynomial. The set of the values of the variables that satisfy this relation is called an implicit curve if $n = 2$ and an implicit surface if $n = 3$. The implicit equations are the basis of algebraic geometry, whose basic subjects of study are the simultaneous solutions of several implicit equations whose left-hand sides are polynomials. These sets of simultaneous solutions are called affine algebraic sets.

In Differential Equations

The solutions of differential equations generally appear expressed by an implicit function.

Applications in Economics

Marginal Rate of Substitution

In economics, when the level set $R(x, y) = 0$ is an indifference curve for the quantities x and y consumed of two goods, the absolute value of the implicit derivative dy/dx is interpreted as the marginal rate of substitution of the two goods: how much more of y one must receive in order to be indifferent to a loss of one unit of x.

Marginal Rate of Technical Substitution

Similarly, sometimes the level set $R(L,K)$ is an isoquant showing various combinations of utilized quantities L of labor and K of physical capital each of which would result in the production of the same given quantity of output of some good. In this case the absolute value of the implicit derivative dK / dL is interpreted as the marginal rate of technical substitution between the two factors of production: how much more capital the firm must use to produce the same amount of output with one less unit of labor.

Optimization

Often in economic theory, some function such as a utility function or a profit function is to be maximized with respect to a choice vector x even though the objective function has not been restricted to any specific functional form. The implicit function theorem guarantees that the first-order conditions of the optimization define an implicit function for each element of the optimal vector x^* of the choice vector x. When profit is being maximized, typically the resulting implicit functions are the labor demand function and the supply functions of various goods. When utility is being maximized, typically the resulting implicit functions are the labor supply function and the demand functions for various goods.

Moreover, the influence of the problem's parameters on x^* —the partial derivatives of the implicit function—can be expressed as total derivatives of the system of first-order conditions found using total differentiation.

Notation for Differentiation

Derivatives are the result of performing a differentiation process upon a function or an expression. Derivative notation is the way we express derivatives mathematically. This is in contrast to natural language where we can simply say "the derivative of...".

There are two main types of notation used to denote the derivative of a function.

Lagrange's Notation is to write the derivative of the function $f(x)$ as

$$f'(x)$$

Leibniz's Notation is to write the derivative of the function f as

$$\frac{df}{dx}$$

Two other notations are worth mentioning

Newton's Notation is to write the derivative of y using a dot

$$\dot{y}$$

Euler's Notation is to use a capital D i.e.

$$D_x f(x)$$

A derivative is always the derivative of a function with respect to a variable. When we write the definition of the derivative as

$$\lim_{\Delta x \to 0} \frac{f(x+\Delta x) - f(x)}{\Delta x}$$

we mean the derivative of the function $f(x)$ with respect to the variable x. One type of notation for derivatives is sometimes called prime notation. The function $f(x)$, which would be read ``$f - prime\ of\ x$", means the derivative of $f(x)$ with respect to x. If we $say\, y = f(x)$, then $y' \left(read\ ``y - prime"\right) = f(x)$. This is even sometimes taken as far as to write things such as, for $y=x^4 + 3x$ (for example), $y' = \left(x^4 + 3x\right)'$. Higher order derivatives in prime notation are represented by increasing the number of primes. For example, the second derivative of y with respect to x would be written as

$$(y')' = y''(x).$$

Beyond the second or third derivative, all those primes get messy, so often the order of the derivative is instead writen as a roman superscript in parenthesis, so that the ninth derivative of $f(x)$ with respect to x is written as $f^{(9)}(x)\ or\ f^{(ix)}(x)$.

A second type of notation for derivatives is sometimes called operator notation. The operator D_x is applied to a function in order to perform differentiation. Then, the derivative of f(x) = y with respect to x can be written as $D_x y$ $(read\ ``D -- sub\ -- x of\ y")$ or as $D_x f(x\ (read\ ``D -- sub x -- of\ -- f(x"))$. Higher order derivatives are written by adding a superscript to D_x, so that, for example the third derivative of $y = (x^2 + sin(x))$ with respect to x would be written as

$$D_x^3\, y(x) = D_x^3\, (x^2 + \sin(x)) = -\cos(x)$$

Another commonly used notation was developed by Leibnitz and is accordingly called Leibnitz notation. With this notation, if $y = f(x)$, then the derivative of y with respect to x can be written as

$$\frac{dy}{dx}$$

(This is read as ``$dy$ -- $dx$", but not ``dy minus dx" or sometimes ``dy over dx"). Since $y = f(x)$, we can also write

$$\frac{df}{dx}$$

This notation suggests that perhaps derivatives can be treated like fractions, which is true in limited ways in some circumstances. (For example with the chain rule.) This is also called differential notation, where dy and dx are differentials. This notation becomes very useful when dealing with differential equations.

A variation of Leibnitz›s differential notation is written instead as

$$\frac{d}{dx}y$$

which resembles the above operator notation, with (d/dx as the operator). Higher order derivatives using leibnitz notation can be written as

$$\left(\frac{d}{dx}\right)^n y(x) = \frac{d^n y}{dx^n}.$$

The exponents may seem to be in strange places in the second form, but it makes sense if you look at the first form.

Leibniz's Notation

The original notation employed by Gottfried Leibniz is used throughout mathematics. It is particularly common when the equation $y = f(x)$ is regarded as a functional relationship between dependent and independent variables y and x. Leibniz's notation makes this relationship explicit by writing the derivative as

$$\frac{dy}{dx}.$$

The function whose value at x is the derivative of f at x is therefore written

$$\frac{df}{dx}(x) \text{ or } \frac{df(x)}{dx} \text{ or } \frac{d}{dx}f(x).$$

Higher derivatives are written as

$$\frac{d^2y}{dx^2}, \frac{d^3y}{dx^3}, \frac{d^4y}{dx^4}, \ldots, \frac{d^ny}{dx^n}.$$

This is a suggestive notational device that comes from formal manipulations of symbols, as in,

$$\frac{d\left(\frac{d\left(\frac{dy}{dx}\right)}{dx}\right)}{dx} = \left(\frac{d}{dx}\right)^3 y = \frac{d^3y}{dx^3}.$$

Logically speaking, these equalities are not theorems. Instead, they are simply definitions of notation.

The value of the derivative of y at a point $x = a$ may be expressed in two ways using Leibniz's notation:

$$\left.\frac{dy}{dx}\right|_{x=a} = \frac{dy}{dx}(a).$$

Leibniz's notation allows one to specify the variable for differentiation (in the denominator). This is especially helpful when considering partial derivatives. It also makes the chain rule easy to remember and recognize:

$$\frac{dy}{dx} = \frac{dy}{du} \cdot \frac{du}{dx}.$$

Leibniz's notation for differentiation does require assigning a meaning to symbols such as dx or dy on their own, and some authors do not attempt to assign these symbols meaning. Leibniz treated these symbols as infinitesimals. Later authors have assigned them other meanings, such as infinitesimals in non-standard analysis or exterior derivatives.

Leibniz's Notation for Antidifferentiation

Leibniz introduced the integral symbol $\int$ in *Analyseos tetragonisticae pars secunda* and *Methodi tangentium inversae exempla* (both from 1675). It is now the standard symbol for integration.

$$\int y'\,dx = \int f'(x)\,dx = f(x) + C_0 = y + C_0$$

$$\int y\,dx = \int f(x)\,dx = F(x) + C_1$$

$$\int\int y\,dx^2 = \int\left(\int y\,dx\right)dx = \int_{X\times X} f(x)\,dx = \int F(x)\,dx = g(x) + C_2$$

$$\underbrace{\int\ldots\int}_{n} y\,\underbrace{dx\ldots dx}_{n} = \int_{\underbrace{X\times\cdots\times X}_{n}} f(x)\,dx = \int s(x)\,dx = S(x) + C_n$$

Lagrange's Notation

One of the most common modern notations for differentiation is due to Joseph Louis Lagrange. In Lagrange's notation, a prime mark denotes a derivative. If f is a function, then its derivative evaluated at x is written

$$f'(x).$$

Lagrange first used the notation in unpublished works, and it appeared in print in 1770.

Higher derivatives are indicated using additional prime marks, as in $f''(x)$ for the second derivative and $f'''(x)$ for the third derivative. The use of repeated prime marks eventually becomes unwieldy. Some authors continue by employing Roman numerals, as in

$$f^{\mathrm{IV}}(x), f^{\mathrm{V}}(x), f^{\mathrm{VI}}(x), \ldots,$$

to denote fourth, fifth, sixth, and higher order derivatives. Other authors use Arabic numerals in parentheses, as in

$$f^{(4)}(x), f^{(5)}(x), f^{(6)}(x), \ldots.$$

This notation also makes it possible to describe the *n*th derivative, where *n* is a variable. This is written

$$f^{(n)}(x).$$

Unicode characters related to Lagrange's notation include

- U+2032 ◌′ PRIME (derivative)
- U+2033 ◌″ DOUBLE PRIME (double derivative)
- U+2034 ◌‴ TRIPLE PRIME (third derivative)
- U+2057 ◌⁗ QUADRUPLE PRIME (fourth derivative)

When there are two independent variables for a function *f*(*x*,*y*), the following convention may be followed:

$$f' = \frac{df}{dx} = f_x$$

$$f_{'} = \frac{df}{dy} = f_y$$

$$f'' = \frac{d^2 f}{dx^2} = f_{xx}$$

$$f_{'}' = \frac{\partial^2 f}{\partial x \partial y} = f_{xy}$$

$$f_{''} = \frac{d^2 f}{dy^2} = f_{yy},$$

Lagrange's Notation for Antidifferentiation

When taking the antiderivative, Lagrange followed Leibniz's notation:

$$f(x) = \int f'(x)dx = \int y\,dx.$$

However, because integration is the inverse of differentiation, Lagrange's notation for higher order derivatives extends to integrals as well. Repeated integrals of *f* may be written as

$f^{-1}(x)$ for the first integral (this is easily confused with the inverse function $f^{-1}(x)$),

$f^{(-2)}(x)$ for the second integral,

$f^{(-3)}(x)$ for the third integral, and

$f^{(-n)}(x)$ for the nth integral.

Euler's Notation

Leonhard Euler's notation uses a differential operator suggested by Louis François Antoine Arbogast, denoted as D (D operator) or D (Newton–Leibniz operator) When applied to a function $f(x)$, it is defined by

$$Df = \frac{df}{dx}.$$

Higher derivatives are notated as powers of D, as in

$D^2 f$ for the second derivative,

$D^3 f$ for the third derivative, and

$D^n f$ for the nth derivative.

Euler's notation leaves implicit the variable with respect to which differentiation is being done. However, this variable can also be notated explicitly. When f is a function of a variable x, this is done by writing

$D_x f$ for the first derivative,

$D_x^2 f$ for the second derivative,

$D_x^3 f$ for the third derivative, and

$D_x^n f$ for the nth derivative.

When f is a function of several variables, it's common to use a "∂" rather than D. As above, the subscripts denote the derivatives that are being taken. For example, the second partial derivatives of a function $f(x, y)$ are:

$$\partial_{xx} f = \frac{\partial^2 f}{\partial x^2},$$

$$\partial_{xy} f = \frac{\partial^2 f}{\partial x \partial y},$$

$$\partial_{yx} f = \frac{\partial^2 f}{\partial y \partial x},$$

$$\partial_{yy} f = \frac{\partial^2 f}{\partial y^2}.$$

Euler's notation is useful for stating and solving linear differential equations, as it simplifies presentation of the differential equation, which can make seeing the essential elements of the problem easier.

Euler's Notation for Antidifferentiation

Euler's notation can be used for antidifferentiation in the same way that Lagrange's notation is. as follows

$D\ \ f(x)$ for a first antiderivative,

$D^{-2} f(x)$ for a second antiderivative, and

$D^{-n} f(x)$ for an nth antiderivative.

Newton's Notation

Newton's notation for differentiation (also called the dot notation for differentiation) places a dot over the dependent variable. That is, if y is a function of t, then the derivative of y with respect to t is

$$\dot{y}$$

Higher derivatives are represented using multiple dots, as in

$$\ddot{y}, \dddot{y}$$

Newton extended this idea quite far:

$$\ddot{y} \equiv \frac{d^2y}{dt^2} = \frac{d}{dt}\left(\frac{dy}{dt}\right) = \frac{d}{dt}\left(\dot{y}\right) = \frac{d}{dt}\left(f'(t)\right) = D_t^2 y = f''(t) = y_t''$$

$$\dddot{y} = \overset{\therefore}{y} \equiv \frac{d^3y}{dt^3} = D_t^3 y = f'''(t) = y_t'''$$

$$\overset{4}{\dot{y}} = \ddddot{y} = \overset{::}{y} \equiv \frac{d^4y}{dt^4} = D_t^4 y = f^{IV}(t) = y_t^{(4)}$$

$$\overset{5}{\dot{y}} = \overset{\cdot\cdot}{\overset{\cdots}{y}} = \overset{\vdots\vdots}{y} = \overset{\because\because}{y} \equiv \frac{d^5y}{dt^5} = D_t^5 y = f^{V}(t) = y_t^{(5)}$$

$$\overset{6}{\dot{y}} = \overset{\cdots}{\overset{\cdots}{y}} \equiv \frac{d^6y}{dt^6} = D_t^6 y = f^{VI}(t) = y_t^{(6)}$$

$$\overset{7}{\dot{y}} = \overset{\cdot}{\overset{\cdots}{\overset{\cdots}{y}}} \equiv \frac{d^7y}{dt^7} = D_t^7 y = f^{VII}(t) = y_t^{(7)}$$

$$\overset{10}{\dot{y}} = \overset{\vdots\vdots}{\overset{\vdots}{y}} \equiv \frac{d^{10}y}{dt^{10}} = D_t^{10} y = f^{X}(t) = y_t^{(10)}$$

$$\overset{n}{\dot{y}} \equiv \frac{d^ny}{dt^n} = D_t^n y = f^{(n)}(t) = y_t^{(n)}$$

Unicode characters related to Newton's notation include:

- U+0307 ◌̇ COMBINING DOT ABOVE (derivative)
- U+0308 ◌̈ COMBINING DIAERESIS (double derivative)
- U+20DB ◌| | COMBINING THREE DOTS ABOVE (third derivative) ← replaced by "combining diaeresis" + "combining dot above".
- U+20DC ◌| | COMBINING FOUR DOTS ABOVE (fourth derivative) ← replaced by "combining diaeresis" twice.
- U+030D ◌̍ COMBINING VERTICAL LINE ABOVE (integral)
- U+030E ◌̎ COMBINING DOUBLE VERTICAL LINE ABOVE (second integral)
- U+25AD ▭ WHITE RECTANGLE (integral)
- U+20DE ◌| | COMBINING ENCLOSING SQUARE (integral)
- U+1DE0 ◌| | COMBINING LATIN SMALL LETTER N (*n*th derivative)

Newton's notation is generally used when the independent variable denotes time. If location y is a function of t, then $\dot{y}$ denotes velocity and $\ddot{y}$ denotes acceleration. This notation is popular in physics and mathematical physics. It also appears in areas of mathematics connected with physics such as differential equations. It is only popular for first and second derivatives, but in applications these are usually the only derivatives that are necessary.

When taking the derivative of a dependent variable $y = f(x)$, an alternative notation exists:

$$\frac{\dot{y}}{\dot{x}} = \dot{y} : \dot{x} \equiv \frac{dy}{dt} : \frac{dx}{dt} = \frac{\frac{dy}{dt}}{\frac{dx}{dt}} = \frac{dy}{dx} = \frac{d}{dt}\big(f(x)\big) = Dy = f'(x) = y'$$

Newton developed the following partial differential operators using side-dots on a curved X (χ). Definitions given by Whiteside are below:

$$\mathcal{X} = f(x, y),$$

$$\cdot\mathcal{X} = x\frac{\partial f}{\partial x} = xf_x,$$

$$\mathcal{X}\cdot = y\frac{\partial f}{\partial y} = yf_y,$$

$$:\mathcal{X} \text{ or } \cdot:\mathcal{X} = x^2\frac{\partial^2 f}{\partial x^2} = x^2 f_{xx},$$

$$\mathcal{X}: \text{ or } \cdot\mathcal{X}\cdot = y^2\frac{\partial^2 f}{\partial y^2} = y^2 f_{yy},$$

$$\cdot\mathcal{X}\cdot \text{ or } \mathcal{X}:\cdot = xy\frac{\partial^2 f}{\partial x \partial y} = xyf_{xy},$$

Newton's Notation for Integration

Newton developed many different notations for integration in his *Quadratura curvarum* (1704) and later works: he wrote a small vertical bar or prime above the dependent variable ($\overset{\prime}{y}$), a prefixing rectangle ($\square y$), or the inclosure of the term in a rectangle ($\boxed{y}$) to denote the *fluent* or time integral (absement).

$$y = \square\dot{y} \equiv \int \dot{y}dt = \int f'(t)dt = D_t^{-1}(D_t y) = f(t) + C_0 = y_t + C_0$$

$$\overset{\prime}{y} = \square y \equiv \int y dt = \int f(t)dt = D_t^{-1} y = F(t) + C_1$$

To denote multiple integrals, Newton used two small vertical bars or primes ($\overset{\prime\prime}{y}$), or a combination of previous symbols $\square\overset{\prime}{y}$ $\boxed{\overset{\prime}{y}}$, to denote the second time integral (absity).

$$\overset{\prime\prime}{y} = \square\overset{\prime}{y} \equiv \int \overset{\prime}{y} dt = \int F(t)dt = D_t^{-2} y = g(t) + C_2$$

Higher order time integrals were as follows:

$$\overset{\prime\prime\prime}{y} = \square\overset{\prime\prime}{y} \equiv \int \overset{\prime\prime}{y} dt = \int g(t)dt = D_t^{-3} y = G(t) + C_3$$

$$\overset{\prime\prime\prime\prime}{y} = \square\overset{\prime\prime\prime}{y} \equiv \int \overset{\prime\prime\prime}{y} dt = \int G(t)dt = D_t^{-4} y = h(t) + C_4$$

$$\overset{\overset{n}{\prime}}{y} = \square\overset{\overset{n-1}{\prime}}{y} \equiv \int \overset{\overset{n-1}{\prime}}{y} dt = \int s(t)dt = D_t^{-n} y = S(t) + C_n$$

This mathematical notation didn't become widespread because of printing difficulties and the Leibniz–Newton calculus controversy.

Partial Derivatives

When more specific types of differentiation are necessary, such as in multivariate calculus or tensor analysis, other notations are common.

For a function $f(x)$, we can express the derivative using subscripts of the independent variable:

$$f_x = \frac{df}{dx}$$

$$f_{xx} = \frac{d^2 f}{dx^2}.$$

This type of notation is especially useful for taking partial derivatives of a function of several variables.

Partial derivatives are generally distinguished from ordinary derivatives by replacing the differential operator d with a "∂" symbol. For example, we can indicate the partial derivative of $f(x, y, z)$ with respect to x, but not to y or z in several ways:

$$\frac{\partial f}{\partial x} = f_x = \partial_x f.$$

Other notations can be found in various subfields of mathematics, physics, and engineering, see for example the Maxwell relations of thermodynamics. The symbol $\left(\frac{\partial T}{\partial V}\right)_S$ is the derivative of the temperature T with respect to the volume V while keeping constant the entropy (subscript) S, while $\left(\frac{\partial T}{\partial V}\right)_P$ is the derivative of the temperature with respect to the volume while keeping constant the pressure P.

Higher-order partial derivatives with respect to one variable are expressed as

$$\frac{\partial^2 f}{\partial x^2} = f_{xx}$$

$$\frac{\partial^3 f}{\partial x^3} = f_{xxx}.$$

Mixed partial derivatives can be expressed as

$$\frac{\partial^2 f}{\partial y \partial x} = f_{xy}.$$

In this last case the variables are written in inverse order between the two notations, explained as follows:

$$(f_x)_y = f_{xy}$$

$$\frac{\partial}{\partial y}\left(\frac{\partial f}{\partial x}\right) = \frac{\partial^2 f}{\partial y \partial x}$$

Notation in Vector Calculus

Vector calculus concerns differentiation and integration of vector or scalar fields. Several notations specific to the case of three-dimensional Euclidean space are common.

Assume that (*x*, *y*, *z*) is a given Cartesian coordinate system, that **A** is a vector field with components $\mathbf{A} = (\mathbf{A}_x, \mathbf{A}_y, \mathbf{A}_z),$, and that $\varphi = \varphi(x, y, z)$ is a scalar field.

The differential operator introduced by William Rowan Hamilton, written ∇ and called del or nabla, is symbolically defined in the form of a vector,

$$\nabla = \left(\frac{\partial}{\partial x}, \frac{\partial}{\partial y}, \frac{\partial}{\partial z}\right),$$

where the terminology *symbolically* reflects that the operator ∇ will also be treated as an ordinary vector.

- Gradient: The gradient $\mathrm{grad}\varphi$ of the scalar field φ is a vector, which is symbolically expressed by the multiplication of ∇ and scalar field φ,

$$\begin{aligned}\mathrm{grad}\,\varphi &= \left(\frac{\partial\varphi}{\partial x},\frac{\partial\varphi}{\partial y},\frac{\partial\varphi}{\partial z}\right)\\ &= \left(\frac{\partial}{\partial x},\frac{\partial}{\partial y},\frac{\partial}{\partial z}\right)\varphi\\ &= \nabla\varphi\end{aligned}$$

- Divergence: The divergence $\mathrm{div}\mathbf{A}$ of the vector field $\mathbf{A}$ is a scalar, which is symbolically expressed by the dot product of ∇ and the vector $\mathbf{A}$,

$$\begin{aligned}\mathrm{div}\,\mathbf{A} &= \frac{\partial A_x}{\partial x}+\frac{\partial A_y}{\partial y}+\frac{\partial A_z}{\partial z}\\ &= \left(\frac{\partial}{\partial x},\frac{\partial}{\partial y},\frac{\partial}{\partial z}\right)\cdot\mathbf{A}\\ &= \nabla\cdot\mathbf{A}\end{aligned}$$

- Laplacian: The Laplacian $\mathrm{div}\,\mathrm{grad}\,\varphi$ of the scalar field φ is a scalar, which is symbolically expressed by the scalar multiplication of ∇^2 and the scalar field φ,

$$\begin{aligned}\mathrm{div}\,\mathrm{grad}\,\varphi &= \nabla\cdot(\nabla\varphi)\\ &= (\nabla\cdot\nabla)\varphi\\ &= \nabla^2\varphi\\ &= \Delta\varphi\end{aligned}$$

- Rotation: The rotation $\mathrm{curl}\mathbf{A}$,, or $\mathrm{rot}\mathbf{A}$,, of the vector field A is a vector, which is symbolically expressed by the cross product of ∇ and the vector A,

$$\begin{aligned}\mathrm{curl}\,\mathbf{A} &= \left(\frac{\partial A_z}{\partial y}-\frac{\partial A_y}{\partial z},\frac{\partial A_x}{\partial z}-\frac{\partial A_z}{\partial x},\frac{\partial A_y}{\partial x}-\frac{\partial A_x}{\partial y}\right)\\ &= \left(\frac{\partial A_z}{\partial y}-\frac{\partial A_y}{\partial z}\right)\mathbf{i}+\left(\frac{\partial A_x}{\partial z}-\frac{\partial A_z}{\partial x}\right)\mathbf{j}+\left(\frac{\partial A_y}{\partial x}-\frac{\partial A_x}{\partial y}\right)\mathbf{k}\\ &= \begin{vmatrix}\mathbf{i} & \mathbf{j} & \mathbf{k}\\ \frac{\partial}{\partial x} & \frac{\partial}{\partial y} & \frac{\partial}{\partial z}\\ A_x & A_y & A_z\end{vmatrix}\\ &= \nabla\times\mathbf{A}\end{aligned}$$

Many symbolic operations of derivatives can be generalized in a straightforward manner by the gradient operator in Cartesian coordinates. For example, the single-variable product rule has a direct analogue in the multiplication of scalar fields by applying the gradient operator, as in

$$(fg)' = f'g + fg' \Rightarrow \nabla(\phi\psi) = (\nabla\phi)\psi + \phi(\nabla\psi).$$

Further notations have been developed for more exotic types of spaces. For calculations in Minkowski space, the d'Alembert operator, also called the d'Alembertian, wave operator, or box operator is represented as $\Box$, or as ∇ when not in conflict with the symbol for the Laplacian.

Differentiation Rules

The operation of *differentiation* or finding the *derivative* of a function has the fundamental property of *linearity*. This property makes taking the derivative easier for functions constructed from the basic elementary functions using the operations of addition and multiplication by a constant number. The basic differentiation rules allow us to compute the derivatives of such functions without using the formal definition of the derivative. Consider these rules in more detail.

Derivative of a Constant

If $f(x) = C$, then

$$f'(x) = C' = 0.$$

In this case, the function $y(x)$ is always equal to to a constant C. Therefore, we can write

$$y(x) = C, y(x+\Delta x) = C.$$

It is clear that the increment of the function is identically equal to zero:

$$\Delta y = y(x+\Delta x) - y(x) = C - C \equiv 0.$$

Substituting this in the limit definition of derivative, we obtain:

$$y'(x) = \lim_{\Delta x \to 0} \frac{\Delta y}{\Delta x} = \lim_{\Delta x \to 0} \frac{y(x+\Delta x) - y(x)}{\Delta x} = \lim_{\Delta x \to 0} \frac{0}{\Delta x} = \lim_{\Delta x \to 0} 0 = 0.$$

Constant Multiple Rule

Let k be a constant. If $f(x)$ is differentiable, then $kf(x)$ is also differentiable and

$$(kf(x))' = kf'(x).$$

Sum Rule

Let $f(x)$ and $g(x)$ be differentiable functions. Then the sum of two functions is also differentiable and

$$(f(x)+g(x))' = f'(x)+g'(x).$$

Let n functions $f_1(x), f_2(x), \ldots, f_n(x)$ be differentiable. Then their sum is also differentiable and

$$[f_1(x)+f_2(x)+\ldots+f_n(x)]' = f_1'(x)+f_2'(x)+\ldots+f_n'(x).$$

Combining the both rules we see that the derivative of difference of two functions is equal to the difference of the derivatives of these functions assuming both of the functions are differentiable:

$$(f(x)-g(x))' = f'(x)-g'(x).$$

We can write the common rule:

Linear Combination Rule

Suppose $f(x)$ and $g(x)$ are differentiable functions and a, b are real numbers. Then the function $h(x) = af(x)+bg(x)$ is also differentiable and

$$h'(x) = af'(x)+bg'(x).$$

We add to this list one more simple rule:

Derivative of the Function $y = x$

If $f(x) = x$, then

$$f'(x) = (x)' = 1.$$

Following the above procedure, we form the ratio $\frac{\Delta y}{\Delta x}$ and find the limit as $\Delta x \to 0$:

$$y'(x) = \lim_{\Delta x \to 0} \frac{\Delta y}{\Delta x} = \lim_{\Delta x \to 0} \frac{(x+\Delta x)-x}{\Delta x} = \lim_{\Delta x \to 0} \frac{\not{x} + \Delta x - \not{x}}{\Delta x}$$

$$= \lim_{\Delta x \to 0} \frac{\Delta \not{x}}{\Delta \not{x}} = \lim_{\Delta x \to 0} 1 = 1.$$

The Product Rule

For the functions *f* and *g*, the derivative of the function $h(x) = f(x)\,g(x)$ with respect to *x* is

$$h'(x) = (f(x)g(x))' = f'(x)g(x)+f(x)g'(x).$$

In Leibniz's notation this is written

$$\frac{d(fg)}{dx} = \frac{df}{dx}g + f\frac{dg}{dx}.$$

The Chain Rule

The derivative of the function $h(x) = f(g(x))$ with respect to x is

$$h'(x) = (f(g(x)))' = f'(g(x)) \cdot g'(x).$$

In Leibniz's notation this is correctly written as:

$$\frac{d}{dx}h(x) = \frac{d}{dz}f(z)|_{z=g(x)} \cdot \frac{d}{dx}g(x),$$

often abridged to $\frac{dh(x)}{dx} = \frac{df(g(x))}{dg(x)} \cdot \frac{dg(x)}{dx}$. Focusing on the notion of maps, and the differential being a map D, this is written in a more concise way as:

$$[\mathrm{D}(h \circ g)]_x = [\mathrm{D}h]_{g(x)} \cdot [\mathrm{D}g]_x.$$

The Inverse Function Rule

If the function *f* has an inverse function *g*, meaning that $g(f(x)) = x$ and $f(g(y)) = y$, then

$$g' = \frac{1}{f' \circ g}.$$

In Leibniz notation, this is written as

$$\frac{dx}{dy} = \frac{1}{\frac{dy}{dx}}.$$

Power Laws, Polynomials, Quotients, and Reciprocals

The Polynomial or Elementary Power Rule

If $f(x) = x^r$, for any real number $r \neq 0$ then

$$f'(x) = rx^{r-1}.$$

Special cases include:

- If $f(x) = x$, then $f'(x) = 1$. This special case may be generalized to:

The derivative of an affine function is constant: if $f(x) = ax + b$, then $f'(x) = a$.

Combining this rule with the linearity of the derivative and the addition rule permits the computation of the derivative of any polynomial.

The Reciprocal Rule

The derivative of $h(x) = 1/f(x)$ for any (nonvanishing) function f is:

$$h'(x) = -\frac{f'(x)}{(f(x))^2}.$$

In Leibniz's notation, this is written

$$\frac{d(1/f)}{dx} = -\frac{1}{f^2}\frac{df}{dx}.$$

The reciprocal rule can be derived from the quotient rule.

The quotient rule

If f and g are functions, then:

$$\left(\frac{f}{g}\right)' = \frac{f'g - g'f}{g^2} \quad \text{wherever g is nonzero.}$$

This can be derived from product rule.

Generalized Power Rule

The elementary power rule generalizes considerably. The most general power rule is the functional power rule: for any functions f and g,

$$(f^g)' = \left(e^{g \ln f}\right)' = f^g\left(f'\frac{g}{f} + g' \ln f\right),$$

wherever both sides are well defined.

Special cases:

- If $f(x) = x^a$, $f'(x) = ax^{a-1}$ when a is any non-zero real number and x is positive.
- The reciprocal rule may be derived as the special case where $g(x) = -1$.

Derivatives of Exponential and Logarithmic Functions

$$\frac{d}{dx}\left(c^{ax}\right) = c^{ax} \ln c \cdot a, \qquad c > 0$$

note that the equation above is true for all c, but the derivative for c < 0 yields a complex number.

$$\frac{d}{dx}\left(e^{ax}\right) = ae^{ax}$$

$$\frac{d}{dx}\left(\log_c x\right) = \frac{1}{x\ln c}, \qquad c > 0, c \neq 1$$

the equation above is also true for all c but yields a complex number if c<0.

$$\frac{d}{dx}\left(\ln x\right) = \frac{1}{x}, \qquad x > 0.$$

$$\frac{d}{dx}\left(\ln|x|\right) = \frac{1}{x}.$$

$$\frac{d}{dx}\left(x^x\right) = x^x(1+\ln x).$$

$$\frac{d}{dx}\left(f(x)^{g(x)}\right) = g(x)f(x)^{g(x)-1}\frac{df}{dx} + f(x)^{g(x)}\ln(f(x))\frac{dg}{dx}, \qquad \text{if } f(x) > 0, \text{ and if } \frac{df}{dx} \text{ and } \frac{dg}{dx} \text{ exist.}$$

$$\frac{d}{dx}\left(f_1(x)^{f_2(x)^{(\ldots)^{f_n(x)}}}\right) = \left[\sum_{k=1}^{n}\frac{\partial}{\partial x_k}\left(f_1(x_1)^{f_2(x_2)^{(\ldots)^{f_n(x_n)}}}\right)\right]\Bigg|_{x_1=x_2=\ldots=x_n=x}, \text{ if } f_{i<n}(x) > 0 \text{ and } \frac{df_i}{dx} \text{ exists.}$$

Logarithmic Derivatives

The logarithmic derivative is another way of stating the rule for differentiating the logarithm of a function (using the chain rule):

$$(\ln f)' = \frac{f'}{f} \qquad \text{wherever } f \text{ is positive.}$$

Logarithmic differentiation is a technique which uses logarithms and its differentiation rules to simplify certain expressions before actually applying the derivative. Logarithms can be used to remove exponents, convert products into sums, and convert division into subtraction, each of which may lead to a simplified expression for taking derivatives.

Derivatives of Trigonometric Functions

$$(\sin x)' = \cos x \qquad\qquad (\arcsin x)' = \frac{1}{\sqrt{1-x^2}}$$

$$(\cos x)' = -\sin x \qquad\qquad (\arccos x)' = -\frac{1}{\sqrt{1-x^2}}$$

$$(\tan x)' = \sec^2 x = \frac{1}{\cos^2 x} = 1 + \tan^2 x \qquad\qquad (\arctan x)' = \frac{1}{1+x^2}$$

$$(\sec x)' = \sec x \tan x \qquad (\operatorname{arcsec} x)' = \frac{1}{|x|\sqrt{x^2-1}}$$

$$(\csc x)' = -\csc x \cot x \qquad (\operatorname{arccsc} x)' = -\frac{1}{|x|\sqrt{x^2-1}}$$

$$(\cot x)' = -\csc^2 x = \frac{-1}{\sin^2 x} = -(1+\cot^2 x) \qquad (\operatorname{arccot} x)' = -\frac{1}{1+x^2}$$

It is common to additionally define an inverse tangent function with two arguments, $\arctan(y, x)$. Its value lies in the range $[-\pi, \pi]$ and reflects the quadrant of the point (x, y). For the first and fourth quadrant (i.e. $x > 0$) one has $\arctan(y, x > 0) = \arctan(y / x)$. Its partial derivatives are

$$\frac{\partial \arctan(y,x)}{\partial y} = \frac{x}{x^2+y^2}, \text{ and } \frac{\partial \arctan(y,x)}{\partial x} = \frac{-y}{x^2+y^2}.$$

Derivatives of Hyperbolic Functions

$$(\sinh x)' = \cosh x = \frac{e^x + e^{-x}}{2} \qquad (\operatorname{arsinh} x)' = \frac{1}{\sqrt{x^2+1}}$$

$$(\cosh x)' = \sinh x = \frac{e^x - e^{-x}}{2} \qquad (\operatorname{arcosh} x)' = \frac{1}{\sqrt{x^2-1}}$$

$$(\tanh x)' = \operatorname{sech}^2 x \qquad (\operatorname{artanh} x)' = \frac{1}{1-x^2}$$

$$(\operatorname{sech} x)' = -\tanh x \operatorname{sech} x \qquad (\operatorname{arsech} x)' = -\frac{1}{x\sqrt{1-x^2}}$$

$$(\operatorname{csch} x)' = -\coth x \operatorname{csch} x \qquad (\operatorname{arcsch} x)' = -\frac{1}{|x|\sqrt{1+x^2}}$$

$$(\coth x)' = -\operatorname{csch}^2 x \qquad (\operatorname{arcoth} x)' = \frac{1}{1-x^2}$$

Derivatives of Special Functions

Gamma function $\Gamma(x) = \int\limits_0^\infty t^{x-1} e^{-t}\, dt$

$$\Gamma'(x) = \int_0^\infty t^{x-1} e^{-t} \ln t\, dt$$

$$= \Gamma(x)\left(\sum_{n=1}^{\infty}\left(\ln\left(1+\frac{1}{n}\right) - \frac{1}{x+n}\right) - \frac{1}{x}\right)$$

$$= \Gamma(x)\psi(x)$$

with $\psi(x)$ being the digamma function, expressed by the parenthesized expression to the right of $\Gamma(x)$ in the line above.

Riemann Zeta function $\zeta(x) = \sum_{n=1}^{\infty} \frac{1}{n^x}$

$$\zeta'(x) = -\sum_{n=1}^{\infty} \frac{\ln n}{n^x} = -\frac{\ln 2}{2^x} - \frac{\ln 3}{3^x} - \frac{\ln 4}{4^x} - \cdots$$

$$= -\sum_{p \text{ prime}} \frac{p^{-x} \ln p}{(1-p^{-x})^2} \prod_{q \text{ prime}, q \neq p} \frac{1}{1-q^{-x}}$$

Derivatives of Integrals

Suppose that it is required to differentiate with respect to x the function

$$F(x) = \int_{a(x)}^{b(x)} f(x,t)dt,$$

where the functions $f(x,t)$ and $\frac{\partial}{\partial x} f(x,t)$ are both continuous in both t and x in some region of the (t,x) plane, including $a(x) \leq t \leq b(x)$, $x_0 \leq x \leq x_1$, and the functions $a(x)$ and $b(x)$ are both continuous and both have continuous derivatives for $x_0 \leq x \leq x_1$. Then for $x_0 \leq x \leq x_1$:

$$F'(x) = f(x,b(x))b'(x) - f(x,a(x))a'(x) + \int_{a(x)}^{b(x)} \frac{\partial}{\partial x} f(x,t)\, dt.$$

This formula is the general form of the Leibniz integral rule and can be derived using the fundamental theorem of calculus.

nth Derivative

A general formula for all of the successive derivatives exists. This formula is called the nth derivative, $f'^n(x)$.

Examples

Calculate the nth derivative of:

$$f(x) = \frac{1}{x}$$

$$f'(x) = -\frac{1}{x^2}$$

$$f''(x) = \frac{1.2}{x^3} = \frac{2!}{x^3}$$

$$f'''(x) = \frac{1\cdot 2\cdot 3}{x^4} = \frac{3!}{x^4}$$

$$\ldots$$

$$f^{n}(x) = (-1)^n \frac{n!}{x^{n+1}}$$

Faà Di Bruno's Formula

If *f* and g are n times differentiable, then

$$\frac{d^n}{dx^n}[f(g(x))] = n!\sum_{\{k_m\}} f^{(r)}(g(x))\prod_{m=1}^{n}\frac{1}{k_m!}\left(g^{(m)}(x)\right)^{k_m}$$

where $r = \sum_{m=1}^{n-1} k_m$ and the set $\{k_m\}$ consists of all non-negative integer solutions of the Diophantine equation

$$\sum_{m=1}^{n} mk_m = n.$$

General Libinitz Rule

The nth derivate of product of 2 functions is given by Leibniz' formula :

$$(fg)^{(n)} = \sum_{k=0}^{n}\binom{n}{k} f^{(n-k)}g^{(k)}$$

where f et g are 2 functions n times derivable, $f^{(l)}$ means l-th derivate of f and $\binom{n}{k} = \frac{n!}{k!(n-k)!}$.

Just like Newton's binomial formula, this formula is easily conjecturable, but much more difficult to prove.

Proof

If you know Newton's binomial formula, you will notice that these 2 formulas (Newton's and Leibniz') are very similar, because they "work" in the same way : induction is the same.

So, we have to use induction with this statement :

$$\forall n \in N \qquad H_n : (fg)^{(n)} = \sum_{k=0}^{n}\binom{n}{k} f^{(n-k)}g^{(k)}$$

that leads us to the following reasoning :

Bases :

$$\text{For} n = 0, (fg)^{(0)} = fg = \binom{0}{0} f^{(0)} g^{(0)} = . So, H_0 holds.$$

Induction steps : For n+1 :

$$(fg)^{(n+1)} = \left((fg)^{(n)}\right)'$$

As we assume H_n holds, we have :

$$(fg)^{(n+1)} = \left(\sum_{k=0}^{n} \binom{n}{k} f^{(n-k)} g^{(k)}\right)' = \sum_{k=0}^{n} \binom{n}{k} f^{(n-k)} g^{(k)}\Bigg)'$$

$$= \sum_{k=0}^{n} \binom{n}{k} \left(f^{(n-k+1)} g^{(k)} + f^{(n-k)} g^{(k+1)}\right)$$

$$= \sum_{k=0}^{n} \binom{n}{k} f^{(n-k+1)} g^{(k)} + \sum_{k=0}^{n} f^{(n-k)} g^{(k+1)}$$

$$= \sum_{k=0}^{n} \binom{n}{k} f^{(n-k+1)} g^{(k)} \sum_{k=1}^{n+1} \binom{n}{k-1} f^{(n-(k-1))} g^{((k-1)+1)}$$

$$= \binom{0}{0} f^{(n+1)} f^{(0)} + \sum_{k=1}^{n} \binom{n}{k} f^{(n+1-k)} g^{(k)} + \binom{n}{n} f^{(0)} g^{(n+1)} \sum_{k=1}^{n} \binom{n}{k-1} f^{(n+1-k)} g^{(k)}$$

$$= \binom{0}{0} f^{(n+1)} g^{(0)} + \binom{n+1}{n+1} f^{(0)} g^{(n+1)} \sum_{k=1}^{n} \left(\binom{n}{k} + \binom{n}{k-1}\right) f^{(n+1-k)} g^{(k)}$$

As Pascal's triangle allows us to see easily that $\binom{n+1}{k} = \binom{n}{k} + \binom{n}{k-1}$, we get :

$$(fg)^{(n+1)} = \binom{0}{0} f^{(n+1)} g^{(0)} + \binom{n+1}{n+1} f^{(0)} g^{(n+1)} + \sum_{k=1}^{n} \binom{n+1}{k} f^{(n+1-k)} g^{(k)}$$

$$= \sum_{k=0}^{n+1} \binom{n+1}{k} f^{(n+1-k)} g^{(k)}$$

Furthermore, $H_n \Rightarrow H_{n+1}$.

Conclusion :

H_n holds $\forall n \in \mathbb{N}$.

We have just proved the formula of nth derivative of product of 2 functions.

References

- Broadbent, T. A. A.; Kline, M. (October 1968). "Reviewed work(s): The History of Ancient Indian Mathematics by C. N. Srinivasiengar". The Mathematical Gazette. 52 (381): 307–8. doi:10.2307/3614212. JSTOR 3614212
- Complex Variables, M.R. Speigel, S. Lipschutz, J.J. Schiller, D. Spellman, Schaum's Outlines Series, McGraw Hill (USA), 2009, ISBN 978-0-07-161569-3
- Weisstein, Eric W. "Double Dot." From MathWorld--A Wolfram Web Resource. "Archived copy". Archived from the original on 2016-03-03. Retrieved 2016-02-05
- Article 580 in Florian Cajori, A History of Mathematical Notations(1929), Dover Publications, Inc. New York. ISBN 0-486-67766-4
- Weisstein, Eric W. "Differential Operator." From MathWorld--A Wolfram Web Resource. "Archived copy". Archived from the original on 2016-01-21. Retrieved 2016-02-07
- Chiang, Alpha C. (1984). Fundamental Methods of Mathematical Economics (Third ed.). New York: McGraw-Hill. ISBN 0-07-010813-7

3

Integral Calculus

Integral calculus is another significant sub-domain of calculus. Integration is the inverse operation of differentiation. This chapter discusses in detail the different kinds of integrals such as, surface integral, multiple integral, Frullani integral, etc. It has been carefully written to provide an easy understanding of the varied aspects of integral calculus.

The branch of mathematics in which the notion of an integral, its properties and methods of calculation are studied. Integral calculus is intimately related to differential calculus, and together with it constitutes the foundation of mathematical analysis. The origin of integral calculus goes back to the early period of development of mathematics and it is related to the method of exhaustion developed by the mathematicians of Ancient Greece (cf. Exhaustion, method of). This method arose in the solution of problems on calculating areas of plane figures and surfaces, volumes of solid bodies, and in the solution of certain problems in statistics and hydrodynamics. It is based on the approximation of the objects under consideration by stepped figures or bodies, composed of simplest planar figures or special bodies (rectangles, parallelopipeds, cylinders, etc.). In this sense, the method of exhaustion can be regarded as an early method of integration. The greatest development of the method of exhaustion in the early period was obtained in the works of Eudoxus (4th century B.C.) and especially Archimedes (3rd century B.C.). Its subsequent application and perfection is associated with the names of several scholars of the 15th–17th centuries.

The fundamental concepts and theory of integral and differential calculus, primarily the relationship between differentiation and integration, as well as their application to the solution of applied problems, were developed in the works of P. de Fermat, I. Newton and G. Leibniz at the end of the 17th century. Their investigations were the beginning of an intensive development of mathematical analysis. The works of L. Euler, Jacob and Johann Bernoulli and J.L. Lagrange played an essential role in its creation in the 18th century. In the 19th century, in connection with the appearance of the notion of a limit, integral calculus achieved a logically complete form (in the works of A.L. Cauchy, B. Riemann and others). The development of the theory and methods of integral calculus took place at the end of 19th century and in the 20th century simultaneously with research into measure theory (cf. Measure), which plays an essential role in integral calculus.

By means of integral calculus it became possible to solve by a unified method many theoretical and applied problems, both new ones which earlier had not been amenable to solution, and old ones that had previously required special artificial techniques. The basic notions of integral calculus are two closely related notions of the integral, namely the indefinite and the definite integral.

The indefinite integral of a given real-valued function on an interval on the real axis is defined as the collection of all its primitives on that interval, that is, functions whose derivatives are the given function. The indefinite integral of a function f is denoted by $\int f(x)dx$.. If F is some primitive of f, then any other primitive of it has the form $F+C$, where C is an arbitrary constant; one therefore writes

$$\int f(x)dx = F(x)+c.$$

The operation of finding an indefinite integral is called integration. Integration is the operation inverse to that of differentiation:

$$\int F^{'}(x)dx = F(x)+C, \quad d\int f(x)dx = f(x)dx$$

The operation of integration is linear: If on some interval the indefinite integrals

$$\int f1(x)dx \; and \int f2(x)dx$$

exist, then for any real numbers $\lambda 1$ and $\lambda 2$, the following integral exists on this interval:

$$\int [\lambda_1 f_1(x)+\lambda_2 f_2(x)]dx$$

and equals

$$\lambda_1 \int f_1(x)dx + \lambda_2 \int f_2(x)dx.$$

For indefinite integrals, the formula of integration by parts holds: If two functions u and are differentiable on some interval and if the integral $\int v\,d\,u$ exists, then so does the integral $\int u\,d\,v$, and the following formula holds:

$$\int u\,d\,v = uv - \int v\,d\,u.$$

The formula for change of variables holds: If for two functions f and ϕ defined on certain intervals, the composite function $f \circ \phi$ makes sense and the function ϕ is differentiable, then the integral

$$\int f[\phi(t)]\,\phi'(t)\,dt$$

exists and equals

$$\int f\,(x)dx$$

A function that is continuous on some bounded interval has a primitive on it and hence an indefinite integral exists for it. The problem of actually finding the indefinite integral of a specified function is complicated by the fact that the indefinite integral of an elementary function is not an elementary function, in general. Many classes of functions are known for which it proves possible to express their indefinite integrals in terms of elementary functions. The simplest examples of these are integrals that are obtained from a table of derivatives of the basic elementary functions:

1) $\int x^{\alpha}dx = \frac{x^{\alpha}+1}{\alpha+1} + C, \alpha \neq -1;$

2) $\int \frac{dx}{x} = In\left|x\right| + C;$

3) $\int a^{x}dx = \frac{a^{x}}{In\, a} + C, a > 0, a \neq 1;$ in particular, $\int e^{x}dx = e^{x} + C;$

4) $\int \sin x dx = -\cos x + C;$

5) $\int \cos x\, d\, x = \sin x + C;$

6) $\int \frac{dx}{\cos^{2} x} = \tan x + C;$

7) $\int \frac{dx}{\sin^{2} x} = -\text{cotan}\ x + C;$

8) $\int \sinh x\, dx = \cosh x + C;$

9) $\int \cosh\ x\ dx = \sinh\ x + C;$

10) $\int \frac{dx}{\cosh^{2} x} = \tanh x + C;$

11) $\frac{dx}{\sinh^{2} x} = -\text{cotanh}\ x + C;$

12) $\int \frac{dx}{x^{2}+a^{2}} = \frac{1}{a}\arctan\frac{x}{a} + C = -\frac{1}{a}\text{arccotan}\frac{x}{a} + C';$

13) $\int \frac{dx}{x^{2}-a^{2}} = \frac{1}{2a} In\left|\frac{x-a}{x+a}\right| + C;$

14) $\int \frac{dx}{\sqrt{a^{2}-x^{2}}} = \arcsin\frac{x}{a} + C = -\arccos\frac{x}{a} + C', \left|x\right| < \left|a\right|;$

15) $\int \frac{dx}{\sqrt{x^{2}\pm a^{2}}} = In\left|x + \sqrt{x^{2}\pm a^{2}}\right| + C($*when* $x^{2}-a^{2}$ is under the square root, it is assumed that $\left|x\right| > \left|a\right|$).

If the denominator of the integrand vanishes at some point, then these formulas are valid only for those intervals inside which the denominator does not vanish.

The indefinite integral of a rational function over any interval on which the denominator does not vanish is a composition of rational functions, arctangents and natural logarithms. Finding the algebraic part of the indefinite integral of a rational function can be achieved by the Ostrogradski method. Integrals of the following types can be reduced by means of substitution and integration by parts to integration of rational functions:

$$\int R\left[x, \left(\frac{ax+b}{ax+b}\right)^{r1}, \ldots, \left(\frac{ax+b}{cx+b}\right)^{rm}\right] dx,$$

where $r1, \ldots, rm$ are rational numbers; integrals of the form

$$\int R\left(x, \sqrt{ax^2+bx+c}\right) dx$$

Certain cases of integrals of differential binomials (cf. Differential binomial; Chebyshev theorem on the integration of binomial differentials); integrals of the form

$$\int R(\sin x, \cos x)dx, \int R(\sin x, \cosh x)dx,$$

(where $R(y1, \ldots, y_n)$ are rational functions); the integrals

$$\int e^{ax} \cos \beta x dx, \int e^{ax} \sin \beta x dx,$$

$$\int x^n \cos \alpha x d x, \qquad \int x^n \sin \alpha x d x,$$

$$\int x^n \arcsin x \, dx \qquad \int x^n arccos \, x \, d x$$

$$\int x^n \arctan x \, d x, \int x^n \operatorname{arccotan} x \, d x, \quad n = 0, 1, \ldots,$$

and many others. In contrast, for example, the integrals

$$\int \frac{e^x}{x^n} dx, \int \frac{\sin x}{x^n} dx, \int \frac{\cos x}{x^n} dx, \quad n = 1, 2, \ldots,$$

cannot be expressed in terms of elementary functions.

The definite integral

$$\int_a^b f(x)dx$$

of a function f defined on an interval $f[a,b]$ is the limit of integral sums of a specific type. If this limit exists, f is said to be Cauchy, Riemann, Lebesgue, etc. integrable.

The geometrical meaning of the integral is tied up with the notion of area: If the function $f \geq 0$ is continuous on the interval $[a,b]$, then the value of the integral

$$\int_a^b f(x)dx$$

is equal to the area of the curvilinear trapezium formed by the graph of the function, that is, the set whose boundary consists of the graph of f, the segment $[a,b]$ and the two segments on the lines

$x = a$ and $x = b$ making the figure closed, which may degenerate to points (cf. Figure).

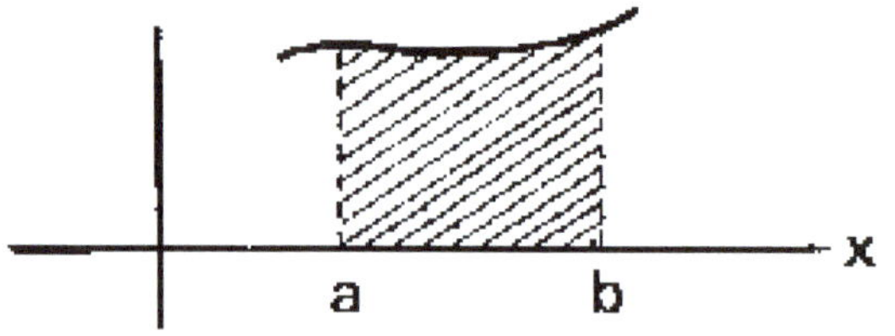

Figure

Formal Definitions

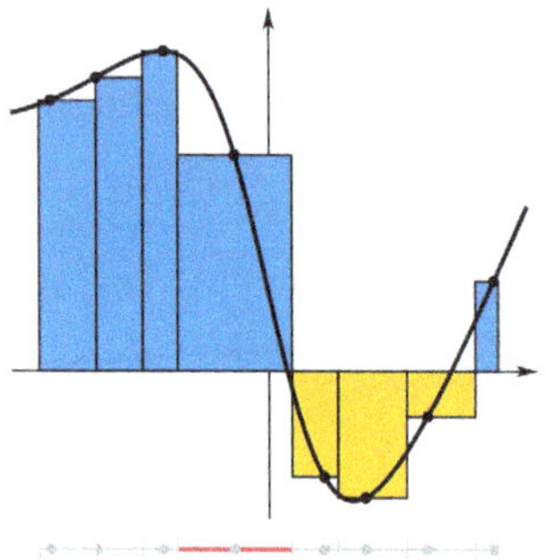

Integral example with irregular partitions (largest marked in red)

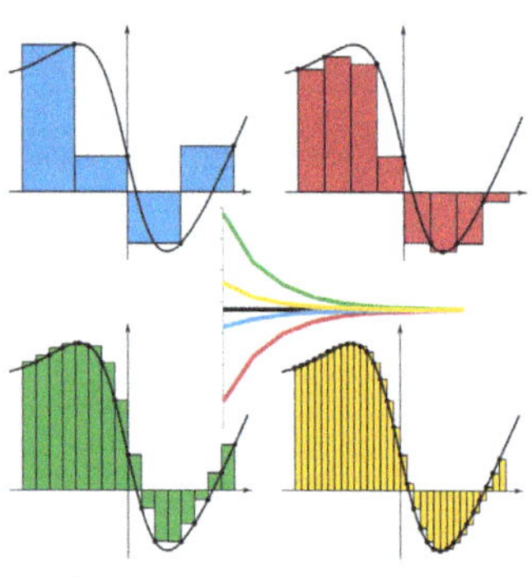

Riemann sums converging

There are many ways of formally defining an integral, not all of which are equivalent. The differences exist mostly to deal with differing special cases which may not be integrable under other definitions, but also occasionally for pedagogical reasons. The most commonly used definitions of integral are Riemann integrals and Lebesgue integrals.

Riemann Integral

The Riemann integral is defined in terms of Riemann sums of functions with respect to *tagged partitions* of an interval. Let $[a, b]$ be a closed interval of the real line; then a *tagged partition* of $[a, b]$ is a finite sequence

$$a = x_0 \leq t_1 \leq x_1 \leq t_2 \leq x_2 \leq \cdots \leq x_{n-1} \leq t_n \leq x_n = b.$$

This partitions the interval $[a, b]$ into n sub-intervals $[x_{i-1}, x_i]$ indexed by i, each of which is"tagged" with a distinguished point $t_i \in [x_{i-1}, x_i]$. A *Riemann sum* of a function f with respect to such a tagged partition is defined as

$$\sum_{i=1}^{n} f(t_i)\Delta_i;$$

thus each term of the sum is the area of a rectangle with height equal to the function value at the distinguished point of the given sub-interval, and width the same as the sub-interval width. Let $\Delta_i = x_i - x_{i-1}$ be the width of sub-interval i; then the *mesh* of such a tagged partition is the width of the largest sub-interval formed by the partition, $\max_{i=1\ldots n} \Delta_i$. The *Riemann integral* of a function f over the interval $[a, b]$ is equal to S if:

For all $\varepsilon > 0$ there exists $\delta > 0$ such that, for any tagged partition $[a, b]$ with mesh less than δ, we have

$$\left| S - \sum_{i=1}^{n} f(t_i)\Delta_i \right| < \varepsilon.$$

When the chosen tags give the maximum (respectively, minimum) value of each interval, the Riemann sum becomes an upper (respectively, lower) Darboux sum, suggesting the close connection between the Riemann integral and the Darboux integral.

Lebesgue Integral

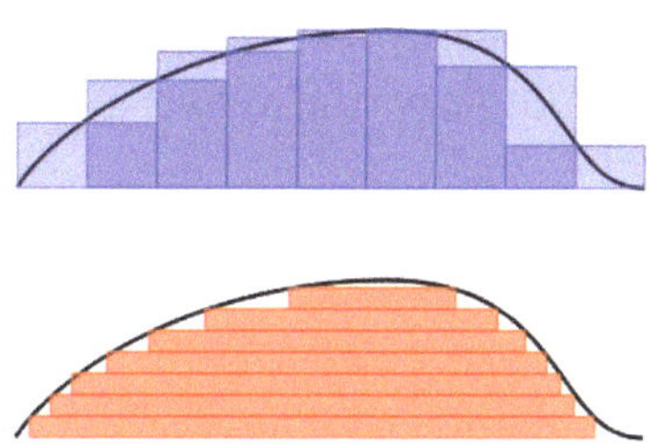

Riemann–Darboux's integration (top) and Lebesgue integration (bottom)

It is often of interest, both in theory and applications, to be able to pass to the limit under the integral. For instance, a sequence of functions can frequently be constructed that approximate, in a suitable sense, the solution to a problem. Then the integral of the solution function should be the limit of the integrals of the approximations. However, many functions that can be obtained as limits are not Riemann-integrable, and so such limit theorems do not hold with the Riemann integral. Therefore, it is of great importance to have a definition of the integral that allows a wider class of functions to be integrated (Rudin 1987).

Such an integral is the Lebesgue integral, that exploits the following fact to enlarge the class of integrable functions: if the values of a function are rearranged over the domain, the integral of a function should remain the same. Thus Henri Lebesgue introduced the integral bearing his name, explaining this integral thus in a letter to Paul Montel:

> I have to pay a certain sum, which I have collected in my pocket. I take the bills and coins out of my pocket and give them to the creditor in the order I find them until I have reached the total sum. This is the Riemann integral. But I can proceed differently. After I have taken

all the money out of my pocket I order the bills and coins according to identical values and then I pay the several heaps one after the other to the creditor. This is my integral.

—*Siegmund-Schultze (2008)*

As Folland (1984) puts it, "To compute the Riemann integral of *f*, one partitions the domain [*a*, *b*] into subintervals", while in the Lebesgue integral, "one is in effect partitioning the range of *f*". The definition of the Lebesgue integral thus begins with a measure, μ. In the simplest case, the Lebesgue measure $\mu(A)$ of an interval $A = [a, b]$ is its width, $b - a$, so that the Lebesgue integral agrees with the (proper) Riemann integral when both exist. In more complicated cases, the sets being measured can be highly fragmented, with no continuity and no resemblance to intervals.

Using the "partitioning the range of *f*" philosophy, the integral of a non-negative function $f : \mathbb{R} \to \mathbb{R}$ should be the sum over t of the areas between a thin horizontal strip between $y = t$ and $y = t + dt$. This area is just $\mu\{x : f(x) > t\}\,dt$. Let $f^*(t) = \mu\{x : f(x) > t\}$. The Lebesgue integral of *f* is then defined by (Lieb & Loss 2001)

$$\int f = \int_0^\infty f^*(t)dt$$

where the integral on the right is an ordinary improper Riemann integral (f^* is a strictly decreasing positive function, and therefore has a well-defined improper Riemann integral). For a suitable class of functions (the measurable functions) this defines the Lebesgue integral.

A general measurable function *f* is Lebesgue-integrable if the sum of the absolute values of the areas of the regions between the graph of *f* and the *x*-axis is finite:

$$\int_E |f|\,d\mu < +\infty.$$

In that case, the integral is, as in the Riemannian case, the difference between the area above the *x*-axis and the area below the *x*-axis:

$$\int_E f\,d\mu = \int_E f^+\,d\mu - \int_E f^-\,d\mu$$

where

$$f^+(x) = \max\{f(x), 0\} = \begin{cases} f(x), & \text{if } f(x) < 0, \\ 0, & \textit{otherwise.} \end{cases}$$

$$f^-(x) = \max\{-f(x), 0\} = \begin{cases} -f(x), & \text{if } f(x) < 0, \\ 0, & \textit{otherwise.} \end{cases}$$

Other Integrals

Although the Riemann and Lebesgue integrals are the most widely used definitions of the integral, a number of others exist, including:

- The Darboux integral, which is constructed using Darboux sums and is equivalent to a Riemann integral, meaning that a function is Darboux-integrable if and only if it is Riemann-integrable. Darboux integrals have the advantage of being simpler to define than Riemann integrals.
- The Riemann–Stieltjes integral, an extension of the Riemann integral.
- The Lebesgue–Stieltjes integral, further developed by Johann Radon, which generalizes the Riemann–Stieltjes and Lebesgue integrals.
- The Daniell integral, which subsumes the Lebesgue integral and Lebesgue–Stieltjes integral without the dependence on measures.
- The Haar integral, used for integration on locally compact topological groups, introduced by Alfréd Haar in 1933.
- The Henstock–Kurzweil integral, variously defined by Arnaud Denjoy, Oskar Perron, and (most elegantly, as the gauge integral) Jaroslav Kurzweil, and developed by Ralph Henstock.
- The Itô integral and Stratonovich integral, which define integration with respect to semimartingales such as Brownian motion.
- The Young integral, which is a kind of Riemann–Stieltjes integral with respect to certain functions of unbounded variation.
- The rough path integral, which is defined for functions equipped with some additional "rough path" structure and generalizes stochastic integration against both semimartingales and processes such as the fractional Brownian motion.

Properties

Linearity

The collection of Riemann-integrable functions on a closed interval $[a, b]$ forms a vector space under the operations of pointwise addition and multiplication by a scalar, and the operation of integration

$$f \mapsto \int_a^b f(x)\, dx$$

is a linear functional on this vector space. Thus, firstly, the collection of integrable functions is closed under taking linear combinations; and, secondly, the integral of a linear combination is the linear combination of the integrals,

$$\int_a^b (\alpha f + \beta g)(x) dx = \alpha \int_a^b f(x) dx + \beta \int_a^b g(x) dx.$$

Similarly, the set of real-valued Lebesgue-integrable functions on a given measure space E with measure μ is closed under taking linear combinations and hence form a vector space, and the Lebesgue integral

$$f \mapsto \int_E f\, d\mu$$

is a linear functional on this vector space, so that

$$\int_E (\alpha f + \beta g)\,d\mu = \alpha \int_E f\,d\mu + \beta \int_E g\,d\mu.$$

More generally, consider the vector space of all measurable functions on a measure space (E, μ), taking values in a locally compact complete topological vector space V over a locally compact topological field K, $f : E \to V$. Then one may define an abstract integration map assigning to each function f an element of V or the symbol ∞,

$$f \mapsto \int_E f\,d\mu,$$

that is compatible with linear combinations. In this situation, the linearity holds for the subspace of functions whose integral is an element of V (i.e. "finite"). The most important special cases arise when K is R, C, or a finite extension of the field Q_p of p-adic numbers, and V is a finite-dimensional vector space over K, and when K = C and V is a complex Hilbert space.

Linearity, together with some natural continuity properties and normalisation for a certain class of "simple" functions, may be used to give an alternative definition of the integral. This is the approach of Daniell for the case of real-valued functions on a set X, generalized by Nicolas Bourbaki to functions with values in a locally compact topological vector space.

Inequalities

A number of general inequalities hold for Riemann-integrable functions defined on a closed and b ounded interval $[a, b]$ and can be generalized to other notions of integral (Lebesgue and Daniell).

- *Upper and lower bounds.* An integrable function f on $[a, b]$, is necessarily bounded on that interval. Thus there are real numbers m and M so that $m \le f(x) \le M$ for all x in $[a, b]$. Since the lower and upper sums of f over $[a, b]$are therefore bounded by, respectively, $m(b - a)$ and $M(b - a)$, it follows that

$$m(b-a) \le \int_a^b f(x)dx \le M(b-a).$$

- *Inequalities between functions.* If $f(x) \le g(x)$ for each x in $[a, b]$ then each of the upper and lower sums of f is bounded above by the upper and lower sums, respectively, of g. Thus

$$\int_a^b f(x)dx \le \int_a^b g(x)dx.$$

This is a generalization of the above inequalities, as $M(b - a)$ is the integral of the constant function with value M over $[a, b]$.

In addition, if the inequality between functions is strict, then the inequality between integrals is also strict. That is, if $f(x) < g(x)$ for each x in $[a, b]$, then

$$\int_a^b f(x)dx < \int_a^b g(x)dx.$$

- *Subintervals.* If $[c, d]$ is a subinterval of $[a, b]$ and $f(x)$ is non-negative for all x, then

  $$\int_c^d f(x)dx \le \int_a^b f(x)dx.$$

- *Products and absolute values of functions.* If f and g are two functions, then we may consider their pointwise products and powers, and absolute values:

  $$(fg)(x) = f(x)g(x),\ f^2(x) = (f(x))^2,\ |f|(x) = |f(x)|.$$

  If f is Riemann-integrable on $[a, b]$ then the same is true for $|f|$, and

  $$\left|\int_a^b f(x)dx\right| \le \int_a^b |f(x)|dx.$$

  Moreover, if f and g are both Riemann-integrable then fg is also Riemann-integrable, and

  $$\left(\int_a^b (fg)(x)dx\right)^2 \le \left(\int_a^b f(x)^2 dx\right)\left(\int_a^b g(x)^2 dx\right).$$

  This inequality, known as the Cauchy–Schwarz inequality, plays a prominent role in Hilbert space theory, where the left hand side is interpreted as the inner product of two square-integrable functions f and g on the interval $[a, b]$.

- *Hölder's inequality.* Suppose that p and q are two real numbers, $1 \le p,\ q \le \infty$ with $1/p + 1/q = 1$, and f and g are two Riemann-integrable functions. Then the functions $|f|^p$ and $|g|^q$ are also integrable and the following Hölder's inequality holds:

  $$\left|\int f(x)g(x)dx\right| \le \left(\int |f(x)|^p dx\right)^{1/p} \left(\int |g(x)|^q dx\right)^{1/q}.$$

  For $p = q = 2$, Hölder's inequality becomes the Cauchy–Schwarz inequality.

- *Minkowski inequality.* Suppose that $p \ge 1$ is a real number and f and g are Riemann-integrable functions. Then $|f|^p$, $|g|^p$ and $|f+g|^p$ are also Riemann-integrable and the following Minkowski inequality holds:

  $$\left(\int |f(x)+g(x)|^p dx\right)^{1/p} \le \left(\int |f(x)|^p dx\right)^{1/p} + \left(\int |g(x)|^p dx\right)^{1/p}.$$

  An analogue of this inequality for Lebesgue integral is used in construction of L^p spaces.

Conventions

In this section, f is a real-valued Riemann-integrable function. The integral

$$\int_a^b f(x)dx$$

over an interval $[a, b]$ is defined if $a < b$. This means that the upper and lower sums of the function f are evaluated on a partition $a = x_0 \le x_1 \le \ldots \le x_n = b$ whose values x_i are increasing. Geometrically, this signifies that integration takes place "left to right", evaluating f within intervals $[x_i, x_{i+1}]$ where an interval with a higher index lies to the right of one with a lower index. The values a and b, the end-points of the interval, are called the limits of integration of f. Integrals can also be defined if $a > b$:

- *Reversing limits of integration.* If $a > b$ then define

  $$\int_a^b f(x)dx = -\int_b^a f(x)dx.$$

  This, with $a = b$, implies:

- *Integrals over intervals of length zero.* If a is a real number then

  $$\int_a^a f(x)dx = 0.$$

  The first convention is necessary in consideration of taking integrals over subintervals of $[a, b]$; the second says that an integral taken over a degenerate interval, or a point, should be zero. One reason for the first convention is that the integrability of f on an interval $[a, b]$ implies that f is integrable on any subinterval $[c, d]$, but in particular integrals have the property that:

- *Additivity of integration on intervals.* If c is any element of $[a, b]$, then

  $$\int_a^b f(x)dx = \int_a^c f(x)dx + \int_c^b f(x)dx.$$

  With the first convention, the resulting relation

  $$\begin{aligned}\int_a^c f(x)dx &= \int_a^b f(x)dx - \int_c^b f(x)dx \\ &= \int_a^b f(x)dx + \int_b^c f(x)dx\end{aligned}$$

  is then well-defined for any cyclic permutation of a, b, and c.

Integrals with a Variable Upper Limit

If a function f is Riemann integrable on an interval $[a,b]$, then the function F defined by

$$F(x) = \int_a^x f(t)\,dt, \quad a \le x \le b,$$

is continuous on this interval. If, in addition, f is continuous at a point $x0$, then F is differentiable at this point and $F'(x_0) = f(x_0)$. In other words, at the points of continuity of a function the following formula holds:

$$\frac{d}{dx} = \int_a^x f(t)\,dt = f(x).$$

Consequently, this formula holds for every Riemann-integrable function on an interval $[a,b]$, except perhaps at a set of points having Lebesgue measure zero, since if a function is Riemann integrable on some interval, then its set of points of discontinuity has measure zero. Thus, if the function f is continuous on $[a,b]$, then the function F defined by

$$F(x)=\int_a^x f(t)\,dt$$

is a primitive of f on this interval. This theorem shows that the operation of differentiation is inverse to that of taking the definite integral with a variable upper limit, and in this way a relationship is established between definite and indefinite integrals:

$$\int f(x)\,dx=\int_a^x f(t)\,dt+c.$$

The geometric meaning of this relationship is that the problem of finding the tangent to a curve and the calculation of the area of plane figures are inverse operations in the above sense.

The following Newton–Leibniz formula holds for any primitive F of an integrable function f on an interval $[a,b]$:

$$\int_a^b f(x)\,dx=F(b)-F(a).$$

It shows that the definite integral of a continuous function over some interval is equal to the difference of the values at the end points of this interval of any primitive of it. This formula is sometimes taken as the definition of the definite integral. Then it is proved that the integral $\int_a^b f(x)\,dx$ introduced in this way is equal to the limit of the corresponding integral sums.

For definite integrals, the formulas for change of variables and integration by parts hold. Suppose, for example, that the function f is continuous on the interval (a,b) and that ϕ is continuous together with its derivative ϕ' on the interval (a,β), where (a,β), is mapped by ϕ into $(a,b): a\langle \varphi(t)\langle b$: for $a\langle t\langle \beta$, so that the composite $f\,o\,\phi$ is meaningful in (a,β), Then, for a_0 , $\beta_0\in(a,\beta)$, the following formulas for change of variables holds:

$$\int_{\phi(a_0)}^{\phi(\beta_0)} f(x)\,dx=\int_{a_0}^{\beta_0} f\left[\phi(t)\right]\phi'(t)\,dt.$$

The formula for integration by parts is:

$$\int_a^b u(x)\,dv(x)=u(x)v(x)\Big|_{x=a}^{x=b}-\int_a^b v(x)\,dv(x),$$

where the functions u and v have Riemann-integrable derivatives on $[a,b]$

Fundamental Theorem of Calculus

The *fundamental theorem of calculus* is the statement that differentiation and integration are inverse operations: if a continuous function is first integrated and then differentiated, the original function is retrieved. An important consequence, sometimes called the *second fundamental theorem of calculus*, allows one to compute integrals by using an antiderivative of the function to be integrated.

Statements of Theorems

Fundamental Theorem of Calculus

Let f be a continuous real-valued function defined on a closed interval $[a, b]$. Let F be the function defined, for all x in $[a, b]$, by

$$F(x) = \int_a^x f(t)dt.$$

Then, F is continuous on $[a, b]$, differentiable on the open interval (a, b), and

$$F'(x) = f(x)$$

for all x in (a, b).

Second Fundamental Theorem of Calculus

Let f be a real-valued function defined on a closed interval $[a, b]$ that admits an antiderivative F on $[a, b]$. That is, f and F are functions such that for all x in $[a, b]$,

$$f(x) = F'(x).$$

If f is integrable on $[a, b]$ then

$$\int_a^b f(x)dx = F(b) - F(a).$$

Calculating Integrals

The second fundamental theorem allows many integrals to be calculated explicitly. For example, to calculate the integral

$$\int_0^1 x^{1/2}\,dx,$$

of the square root function $f(x) = x^{1/2}$ between 0 and 1, it is sufficient to find an antiderivative, that is, a function $F(x)$ whose derivative equals $f(x)$:

$$F'(x) = f(x).$$

One such function is $F(x) = 2/3x^{3/2}$. Then the value of the integral in question is

$$\int_0^1 x^{1/2}\,dx = F(1) - F(0) = \frac{2}{3}(1)^{3/2} - \frac{2}{3}(0)^{3/2} = \frac{2}{3}.$$

This is a case of a general rule, that for $f(x) = x^q$, with $q \neq -1$, the related function, the so-called antiderivative is $F(x) = x^{q+1}/(q+1)$. Tables of this and similar antiderivatives can be used to calculate integrals explicitly, in much the same way that tables of derivatives can be used.

Extensions

Improper Integrals

A "proper" Riemann integral assumes the integrand is defined and finite on a closed and bounded interval, bracketed by the limits of integration. An improper integral occurs when one or more of these conditions is not satisfied. In some cases such integrals may be defined by considering the limit of a sequence of proper Riemann integrals on progressively larger intervals.

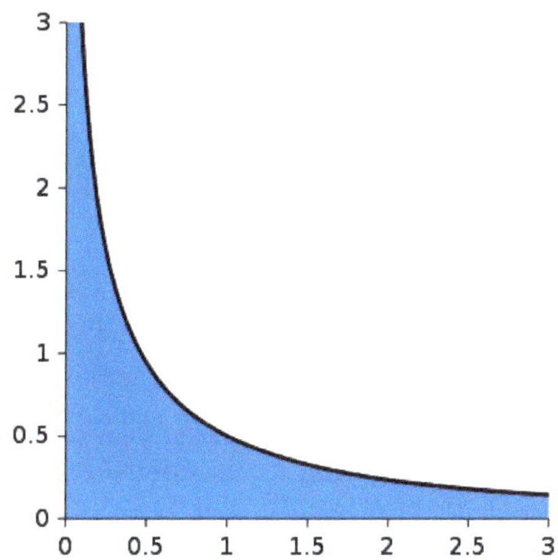

The improper integral $\int_0^\infty \frac{dx}{(x+1)\sqrt{x}} = \pi$ has unbounded intervals for both domain and range.

If the interval is unbounded, for instance at its upper end, then the improper integral is the limit as that endpoint goes to infinity.

$$\int_a^\infty f(x)dx = \lim_{b\to\infty} \int_a^b f(x)dx$$

If the integrand is only defined or finite on a half-open interval, for instance $(a, b]$, then again a limit may provide a finite result.

$$\int_a^b f(x)dx = \lim_{\epsilon\to 0} \int_{a+\epsilon}^b f(x)dx$$

That is, the improper integral is the limit of proper integrals as one endpoint of the interval of integration approaches either a specified real number, or ∞, or $-\infty$. In more complicated cases, limits are required at both endpoints, or at interior points.

Multiple Integration

Double integral as volume under a surface

Just as the definite integral of a positive function of one variable represents the area of the region between the graph of the function and the x-axis, the *double integral* of a positive function of two variables represents the volume of the region between the surface defined by the function and the plane that contains its domain. For example, a function in two dimensions depends on two real variables, x and y, and the integral of a function f over the rectangle R given as the Cartesian product of two intervals $R=[a,b]\times[c,d]$ can be written

$$\int_R f(x,y)dA$$

where the differential dA indicates that integration is taken with respect to area. This double integral can be defined using Riemann sums, and represents the (signed) volume under the graph of $z=f(x,y)$ over the domain R. Under suitable conditions (e.g., if f is continuous), then Fubini's theorem guarantees that this integral can be expressed as an equivalent iterated integral

$$\int_a^b\left[\int_c^d f(x,y)dy\right]dx.$$

This reduces the problem of computing a double integral to computing one-dimensional integrals. Because of this, another notation for the integral over R uses a double integral sign:

$$\iint_R f(x,y)dA.$$

Integration over more general domains is possible. The integral of a function f, with respect to volume, over a subset D of R^n is denoted by notation such as

$$\int_D f(\mathbf{x})d^n\mathbf{x},\quad \int_D f\,dV$$

or similar.

Line Integrals

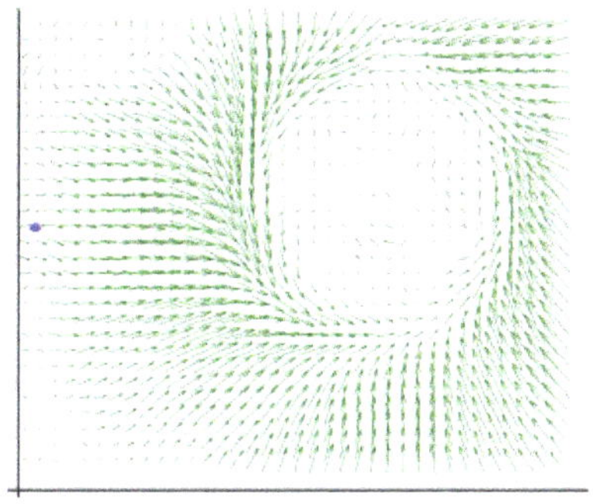

A line integral sums together elements along a curve.

The concept of an integral can be extended to more general domains of integration, such as curved lines and surfaces. Such integrals are known as line integrals and surface integrals respectively. These have important applications in physics, as when dealing with vector fields.

A *line integral* (sometimes called a *path integral*) is an integral where the function to be integrated is evaluated along a curve. Various different line integrals are in use. In the case of a closed curve it is also called a *contour integral*.

The function to be integrated may be a scalar field or a vector field. The value of the line integral is the sum of values of the field at all points on the curve, weighted by some scalar function on the curve (commonly arc length or, for a vector field, the scalar product of the vector field with a differential vector in the curve). This weighting distinguishes the line integral from simpler integrals defined on intervals. Many simple formulas in physics have natural continuous analogs in terms of line integrals; for example, the fact that work is equal to force, **F**, multiplied by displacement, **s**, may be expressed (in terms of vector quantities) as:

$$W = \mathbf{F} \cdot \mathbf{s}$$

For an object moving along a path C in a vector field **F** such as an electric field or gravitational field, the total work done by the field on the object is obtained by summing up the differential work done in moving from **s** to **s** + d**s**. This gives the line integral

$$W = \int_C \mathbf{F} \cdot d\mathbf{s}.$$

Surface Integrals

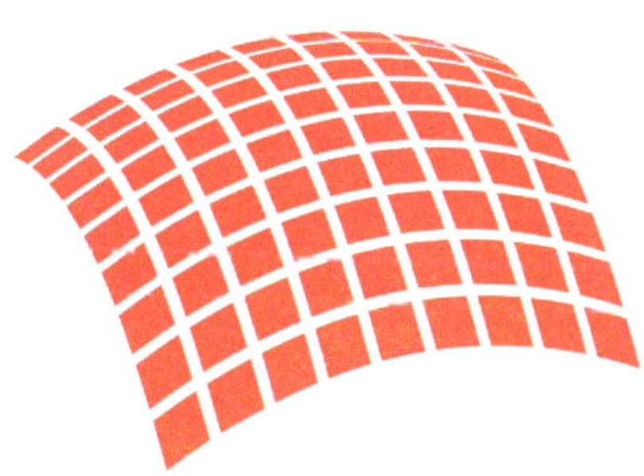

The definition of surface integral relies on splitting the surface into small surface elements.

A *surface integral* is a definite integral taken over a surface (which may be a curved set in space); it can be thought of as the double integral analog of the line integral. The function to be integrated may be a scalar field or a vector field. The value of the surface integral is the sum of the field at all points on the surface. This can be achieved by splitting the surface into surface elements, which provide the partitioning for Riemann sums.

For an example of applications of surface integrals, consider a vector field **v** on a surface S; that is, for each point x in S, $\mathbf{v}(x)$ is a vector. Imagine that we have a fluid flowing through S, such that $\mathbf{v}(x)$ determines the velocity of the fluid at x. The flux is defined as the quantity of fluid flowing through S in unit amount of time. To find the flux, we need to take the dot product of **v** with the unit surface normal to S at each point, which will give us a scalar field, which we integrate over the surface:

$$\int_S \mathbf{v} \cdot d\mathbf{S}.$$

The fluid flux in this example may be from a physical fluid such as water or air, or from electrical or magnetic flux. Thus surface integrals have applications in physics, particularly with the classical theory of electromagnetism.

Contour Integrals

In complex analysis, the integrand is a complex-valued function of a complex variable z instead of a real function of a real variable x. When a complex function is integrated along a curve γ in the complex plane, the integral is denoted as follows

$$\int_\gamma f(z)dz$$

This is known as a contour integral.

Integrals of Differential Forms

A differential form is a mathematical concept in the fields of multivariable calculus, differential topology, and tensors. Differential forms are organized by degree. For example, a one-form is a weighted sum of the differentials of the coordinates, such as:

$$E(x,y,z)dx + F(x,y,z)dy + G(x,y,z)dz.$$

where E, F, G are functions in three dimensions. A differential one-form can be integrated over an oriented path, and the resulting integral is just another way of writing a line integral. Here the basic differentials dx, dy, dz measure infinitesimal oriented lengths parallel to the three coordinate axes.

A differential two-form is a sum of the form

$$G(x,y,z)dx \wedge dy + E(x,y,z)dy \wedge dz + F(x,y,z)dz \wedge dx.$$

Here the basic two-forms $dx \wedge dy, dz \wedge dx, dy \wedge dz.$ measure oriented areas parallel to the coordinate two-planes. The symbol $\wedge$ denotes the wedge product, which is similar to the cross product in the sense that the wedge product of two forms representing oriented lengths represents an oriented area. A two-form can be integrated over an oriented surface, and the resulting integral is equivalent to the surface integral giving the flux of $E\mathbf{i}+F\mathbf{j}+G\mathbf{k}$.

Unlike the cross product, and the three-dimensional vector calculus, the wedge product and the calculus of differential forms makes sense in arbitrary dimension and on more general manifolds (curves, surfaces, and their higher-dimensional analogs). The exterior derivative plays the role of the gradient and curl of vector calculus, and Stokes' theorem simultaneously generalizes the three theorems of vector calculus: the divergence theorem, Green's theorem, and the Kelvin-Stokes' theorem.

Summations

The discrete equivalent of integration is summation. Summations and integrals can be put on the same foundations using the theory of Lebesgue integrals or time scale calculus.

Computation

Analytical

The most basic technique for computing definite integrals of one real variable is based on the fundamental theorem of calculus. Let $f(x)$ be the function of x to be integrated over a given interval $[a, b]$. Then, find an antiderivative of f; that is, a function F such that $F' = f$ on the interval. Provided the integrand and integral have no singularities on the path of integration, by the fundamental theorem of calculus,

$$\int_a^b f(x)dx = F(b) - F(a).$$

The integral is not actually the antiderivative, but the fundamental theorem provides a way to use antiderivatives to evaluate definite integrals.

The most difficult step is usually to find the antiderivative of f. It is rarely possible to glance at a function and write down its antiderivative. More often, it is necessary to use one of the many techniques that have been developed to evaluate integrals. Most of these techniques rewrite one integral as a different one which is hopefully more tractable. Techniques include:

- Integration by substitution
- Integration by parts
- Inverse function integration
- Changing the order of integration
- Integration by trigonometric substitution
- Tangent half-angle substitution
- Integration by partial fractions
- Integration by reduction formulae

- Integration using parametric derivatives
- Integration using Euler's formula
- Euler substitution
- Differentiation under the integral sign
- Contour integration

Alternative methods exist to compute more complex integrals. Many nonelementary integrals can be expanded in a Taylor series and integrated term by term. Occasionally, the resulting infinite series can be summed analytically. The method of convolution using Meijer G-functions can also be used, assuming that the integrand can be written as a product of Meijer G-functions. There are also many less common ways of calculating definite integrals; for instance, Parseval's identity can be used to transform an integral over a rectangular region into an infinite sum. Occasionally, an integral can be evaluated by a trick.

Computations of volumes of solids of revolution can usually be done with disk integration or shell integration.

Symbolic

Many problems in mathematics, physics, and engineering involve integration where an explicit formula for the integral is desired. Extensive tables of integrals have been compiled and published over the years for this purpose. With the spread of computers, many professionals, educators, and students have turned to computer algebra systems that are specifically designed to perform difficult or tedious tasks, including integration. Symbolic integration has been one of the motivations for the development of the first such systems, like Macsyma.

A major mathematical difficulty in symbolic integration is that in many cases, a closed formula for the antiderivative of a rather simple-looking function does not exist. For instance, it is known that the antiderivatives of the functions $\exp(x^2)$, x^x and $(\sin x)/x$ cannot be expressed in the closed form involving only rational and exponential functions, logarithm, trigonometric functions and inverse trigonometric functions, and the operations of multiplication and composition; in other words, none of the three given functions is integrable in elementary functions, which are the functions which may be built from rational functions, roots of a polynomial, logarithm, and exponential functions. The Risch algorithm provides a general criterion to determine whether the antiderivative of an elementary function is elementary, and, if it is, to compute it. Unfortunately, it turns out that functions with closed expressions of antiderivatives are the exception rather than the rule. Consequently, computerized algebra systems have no hope of being able to find an antiderivative for a randomly constructed elementary function. On the positive side, if the 'building blocks' for antiderivatives are fixed in advance, it may be still be possible to decide whether the antiderivative of a given function can be expressed using these blocks and operations of multiplication and composition, and to find the symbolic answer whenever it exists. The Risch algorithm, implemented in Mathematica and other computer algebra systems, does just that for functions and antiderivatives built from rational functions, radicals, logarithm, and exponential functions.

Some special integrands occur often enough to warrant special study. In particular, it may be useful to have, in the set of antiderivatives, the special functions (like the Legendre functions, the hypergeometric function, the gamma function, the incomplete gamma function and so on.) Extending the Risch's algorithm to include such functions is possible but challenging and has been an active research subject.

More recently a new approach has emerged, using *D*-finite functions, which are the solutions of linear differential equations with polynomial coefficients. Most of the elementary and special functions are *D*-finite, and the integral of a *D*-finite function is also a *D*-finite function. This provides an algorithm to express the antiderivative of a *D*-finite function as the solution of a differential equation.

This theory also allows one to compute the definite integral of a *D*-function as the sum of a series given by the first coefficients, and provides an algorithm to compute any coefficient.[

Numerical

Some integrals found in real applications can be computed by closed-form antiderivatives. Others are not so accommodating. Some antiderivatives do not have closed forms, some closed forms require special functions that themselves are a challenge to compute, and others are so complex that finding the exact answer is too slow. This motivates the study and application of numerical approximations of integrals. This subject, called *numerical integration* or *numerical quadrature*, arose early in the study of integration for the purpose of making hand calculations. The development of general-purpose computers made numerical integration more practical and drove a desire for improvements. The goals of numerical integration are accuracy, reliability, efficiency, and generality, and sophisticated modern methods can vastly outperform a naive method by all four measures (Dahlquist & Björck 2008; Kahaner, Moler & Nash 1989; Stoer & Bulirsch 2002).

Consider, for example, the integral

$$\int_{-2}^{2} \frac{1}{5}\left(\frac{1}{100}(322+3x(98+x(37+x)))-24\frac{x}{1+x^2}\right)dx$$

which has the exact answer 94/25 = 3.76. (In ordinary practice, the answer is not known in advance, so an important task — not explored here — is to decide when an approximation is good enough.) A "calculus book" approach divides the integration range into, say, 16 equal pieces, and computes function values.

Spaced function values									
x	−2.00	−1.50	−1.00	−0.50	0.00	0.50	1.00	1.50	2.00
***f*(*x*)**	2.22800	2.45663	2.67200	2.32475	0.64400	−0.92575	−0.94000	−0.16963	0.83600

x		−1.75	−1.25	−0.75	−0.25	0.25	0.75	1.25	1.75	
***f*(*x*)**		2.33041	2.58562	2.62934	1.64019	−0.32444	−1.09159	−0.60387	0.31734	

Using the left end of each piece, the rectangle method sums 16 function values and multiplies by the step width, h, here 0.25, to get an approximate value of 3.94325 for the integral. The accuracy is not impressive, but calculus formally uses pieces of infinitesimal width, so initially this may seem little cause for concern. Indeed, repeatedly doubling the number of steps eventually produces an approximation of 3.76001. However, 2^{18} pieces are required, a great computational expense for such little accuracy; and a reach for greater accuracy can force steps so small that arithmetic precision becomes an obstacle.

A better approach replaces the rectangles used in a Riemann sum with trapezoids. The trapezoid rule is almost as easy to calculate; it sums all 17 function values, but weights the first and last by one half, and again multiplies by the step width. This immediately improves the approximation to 3.76925, which is noticeably more accurate. Furthermore, only 2^{10} pieces are needed to achieve 3.76000, substantially less computation than the rectangle method for comparable accuracy. The idea behind the trapezoid rule, that more accurate approximations to the function yield better approximations to the integral, can be carried further. Simpson's rule approximates the integrand by a piecewise quadratic function. Riemann sums, the trapezoid rule, and Simpson's rule are examples of a family of quadrature rules called Newton–Cotes formulas. The degree n Newton–Cotes quadrature rule approximates the polynomial on each subinterval by a degree n polynomial. This polynomial is chosen to interpolate the values of the function on the interval. Higher degree Newton-Cotes approximations can be more accurate, but they require more function evaluations (already Simpson's rule requires twice the function evaluations of the trapezoid rule), and they can suffer from numerical inaccuracy due to Runge's phenomenon. One solution to this problem is Clenshaw–Curtis quadrature, in which the integrand is approximated by expanding it in terms of Chebyshev polynomials. This produces an approximation whose values never deviate far from those of the original function.

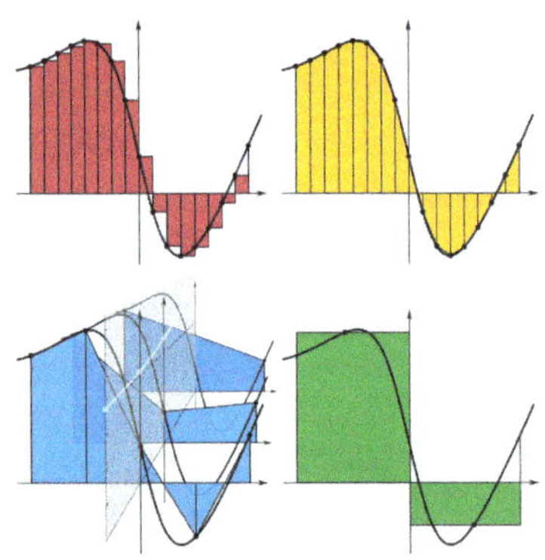

Numerical quadrature methods: Rectangle, Trapezoid, Romberg, Gauss

Romberg's method builds on the trapezoid method to great effect. First, the step lengths are halved incrementally, giving trapezoid approximations denoted by $T(h_0)$, $T(h_1)$, and so on, where h_{k+1} is half of h_k. For each new step size, only half the new function values need to be computed; the others carry over from the previous size (as shown in the table above). But the really powerful idea is to interpolate a polynomial through the approximations, and extrapolate to $T(0)$. With this method a numerically *exact* answer here requires only four pieces (five function values). The Lagrange polynomial interpolating $\{h_k,T(h_k)\}_{k=0...2} = \{(4.00,6.128), (2.00,4.352), (1.00,3.908)\}$ is $3.76 + 0.148h^2$, producing the extrapolated value 3.76 at $h = 0$.

Gaussian quadrature often requires noticeably less work for superior accuracy. In this example, it can compute the function values at just two x positions, $\pm 2/\sqrt{3}$, then double each value and sum to get the numerically exact answer. The explanation for this dramatic success lies in the choice of points. Unlike Newton–Cotes rules, which interpolate the integrand at evenly spaced points, Gaussian quadrature evaluates the function at the roots of a set of orthogonal polynomials. An n-point Gaussian method is exact for polynomials of degree up to $2n - 1$. The function in this example is a degree 3 polynomial, plus a term that cancels because the chosen endpoints are symmetric around zero. (Cancellation also benefits the Romberg method.)

In practice, each method must use extra evaluations to ensure an error bound on an unknown function; this tends to offset some of the advantage of the pure Gaussian method, and motivates the popular Gauss–Kronrod quadrature formulae. More broadly, adaptive quadrature partitions a range into pieces based on function properties, so that data points are concentrated where they are needed most.

The computation of higher-dimensional integrals (for example, volume calculations) makes important use of such alternatives as Monte Carlo integration.

A calculus text is no substitute for numerical analysis, but the reverse is also true. Even the best adaptive numerical code sometimes requires a user to help with the more demanding integrals. For example, improper integrals may require a change of variable or methods that can avoid infinite function values, and known properties like symmetry and periodicity may provide critical leverage. For example, the integral $\int x^{-1/2}e^{-} \, dx$ is difficult to evaluate numerically because it is infinite at $x = 0$. However, the substitution $u = \sqrt{x}$ transforms the integral into $2\int_0^1 e^{-u^2} \, du$, which has no singularities at all.

Multiple Integral

A Multiple Integral Is A generalization of the usual integral in one dimension to functions of multiple variables in higher-dimensional spaces, e.g.

$$\iint f(x,y) \, dx \, dy,$$

which is an integral of a function over a two-dimensional region. The most common multiple integrals are double and triple integrals, involving two or three variables, respectively. Since the world has three spatial dimensions, many of the fundamental equations of physics involve multiple integration (e.g. with respect to each spatial variable).

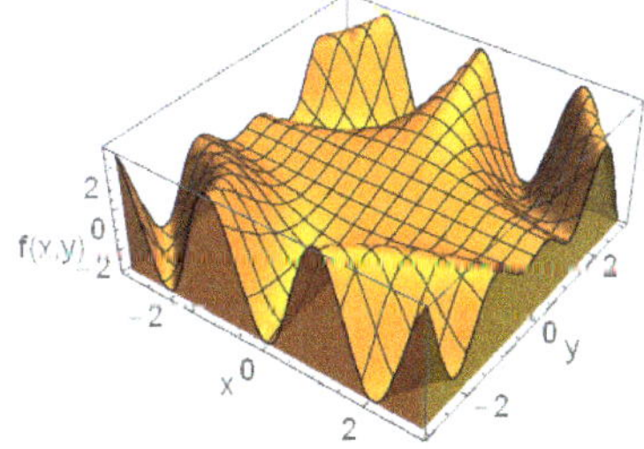

The double integral of the graphed function corresponds to the volume contained underneath the surface corresponding to the function.

Often, multiple integrals are written with a single integral symbol but the notation still implies the correct number of integrals, as in some of the following:

$$\int f\, dA,\ \int f\, d^3x,\ \int f(x_1,\dots\dots,x_n)\, dx_1 \dots dx_n,\ \int_{S_2} f,$$

referring to integration over an area A, all of $\mathbb{R}^3$, all of $\mathbb{R}^n$, and the surface of the sphere S^2, respectively. Multiple integrals are powerful tools because they allow us to do all the things that can be done in one dimension by integrating, like finding average values or work done, in multiple dimensions.

Double Integrals

As in partial differentiation where all but one variable is treated as constant when a partial derivative is taken, in multiple integration all but one variable is treated as constant and integrals are taken iteratively. That is, given an integral

$$\int\int f(x,y)\, dy\, dx,$$

one should treat x as a constant and perform the integration with respect to y, then perform the integration with respect to x. Note that the integral limits for y may be functions of x in this case; these correspond to performing the integration over some region that is not a rectangle in the xy -plane.

According to Fubini's theorem, in most cases one can interchange the order of integration between x and y as desired. This is often very useful, as some integrals can only be evaluated easily in one order. Fubini's theorem fails, however, when $\int\int |f(x,y)|\, dydx = \infty$.

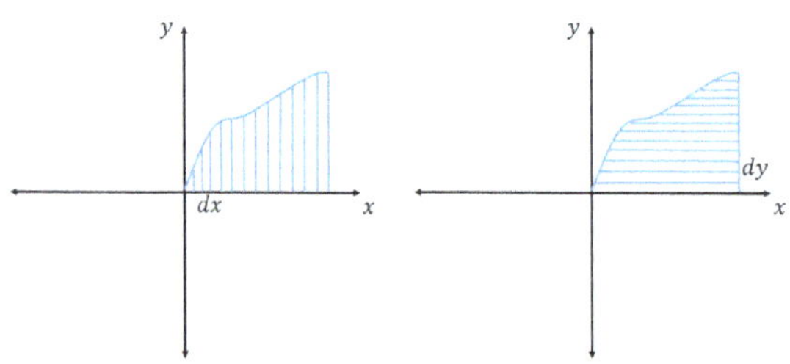

Integration over the same region in 2D can be "sliced" differently by performing or integration first.

Geometrically, a double integral corresponds to the volume under some surface in $\mathbb{R}^3$. The double integral of the function $f(x,y)=1$ therefore corresponds to the area of the region of integration.

Example

Compute the double integral

$$\int_0^1\int_0^1 \frac{x^2+y^2}{2}\, dx\, dy.$$

First perform the integration with respect to x, and then with respect to y:

$$\int_0^1\int_0^1 \frac{x^2+y^2}{2}\, dx\, dy = \int_0^1 \left(\frac{x^3}{6}+\frac{y^2x}{2}\right)\Bigg|_0^1 dy = \int_0^1 \frac{1}{6}+\frac{y^2}{2}\, dy = \frac{y+y^3}{6}\Bigg|_0^1 = \frac{1}{3}.$$

This integral geometrically corresponds to a sum of the given function over the square with side lengths1 .

Example

Evaluate the triple integral

$$\int_0^2 \int_{-3}^0 \int_{-1}^1 \left(x^2 + yz\right) dz\, dy\, dx.$$

Since z is the innermost differential, we integrate with respect to that first. We need to treat all other variables as constants, and integrate with respect to z. This will go as follows:

$$\int_0^2 \int_{-3}^0 \left[x^2\, z + \frac{1}{2} yz^2 \right]_{z=-1}^{z=1} dy\, dx.$$

Just like any other integral, we will plug in the top bound plugged in for z and then subtract the bottom bound plugged in for z , and get

$$\int_0^2 \int_{-3}^0 \left[x^2\,(1) + \frac{1}{2} y\,(1^2) - \left(x^2\,(-1) + \frac{1}{2} y\,(-1^2) \right) \right] dy\, dx.$$

This simplifies to

$$\int_0^2 \int_{-3}^0 2x^2\; dy\, dx = \int_0^2 6x^2 dx = 16.$$

Example

Compute the double integral

$$\int_0^1 \int_y^1 e^{-x^2}\; dxdy.$$

Since e^{-x^2} is always positive and finite, and the integration region is a finite area, the integral of the absolute value is finite, so Fubini's theorem applies: the order of integration may be exchanged. This is ideal because the Gaussian integral cannot be computed directly. Note that the integration is over the region:

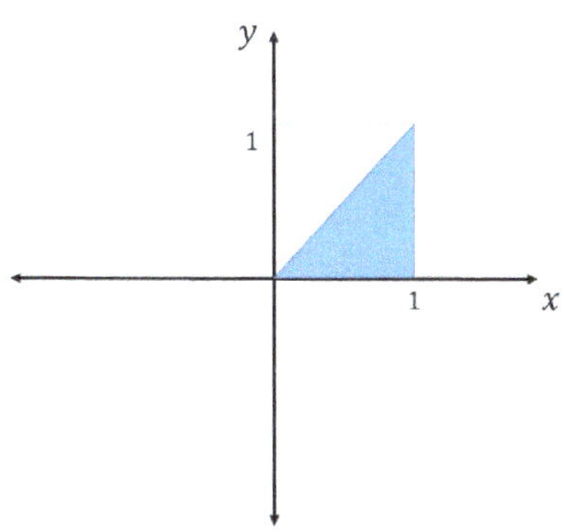

Region of integration is the triangle given in blue.

Changing the order of integration, the integral becomes

$$\int_0^1 \int_0^x e^{-x^2}\, dxdy = \int_0^1 xe^{-x^2} dx = \frac{1}{2}\left(1-e^{-1}\right).$$

Note the change in the integral limits coming from the change of variables; after the change of variables, the integration is first performed up from the x-axis $y=0$ to the line $y=x$, and then across from $x=0$ to $x=1$.

Example

Compute the double integrals

$$\int_0^1 \int_0^1 \frac{x^2-y^2}{\left(x^2+y^2\right)^2}\, dx\, dy, \quad \int_0^1 \int_0^1 \frac{x^2-y^2}{\left(x^2+y^2\right)^2}\, dx\, dy.$$

Conveniently, the integrand can be written as a mixed second partial derivative:

$$\frac{x^2-y^2}{\left(x^2+y^2\right)^2} = -\frac{\partial^2}{\partial x \partial y} \arctan\left(y/x\right).$$

A careful analysis of the values of the partial derivatives at the points $(0,1)$ and the origin gives the first double integral

$$\begin{aligned}\int_0^1 \int_0^1 \frac{x^2-y^2}{\left(x^2+y^2\right)^2}\, dx\, dy &= \int_0^1 \left(-\frac{\partial^2}{\partial y} \arctan\left(y/x\right)\Big|_{x=1} + \frac{\partial^2}{\partial y} \arctan\left(y/x\right)\Big|_{x=0} \right) dy \\ &= \int_0^1 \left(-\frac{\partial}{\partial y} \arctan\left(y\right) + \frac{\partial}{\partial y} \arctan\left(\infty\right) \right) dy \\ &= -\arctan\left(1\right) = -\frac{\pi}{4}.\end{aligned}$$

Performing the other integral yields by symmetry (note the negative sign in the numerator of the integrand)

$$\int_0^1 \int_0^1 \frac{x^2-y^2}{\left(x^2+y^2\right)^2}\, dydx = \frac{\pi}{4}.$$

There is a factor of -1 discrepancy between the two different orderings of the integral. This is because the conditions for Fubini's theorem is violated. Near the origin, $\left(x^2+y^2\right)^2$ diverges rapidly and so the integral $\int_0^1 \int_0^1 \left| \frac{x^2-y^2}{\left(x^2+y^2\right)^2} \right| dydx$ diverges.

For certain functions, it may be advantageous to change the coordinate system used in evaluating integrals. For instance, polar coordinates $x = r\cos\theta,\ y = r\sin\theta$ may be used. In this case, however, one must be careful, because the area element dA is different in Cartesian and polar coordinates. Specifically,

$$\iint f(x,y)\,dxdy = \iint f(r\cos\theta,\ r\sin\theta)\left|\frac{\partial(x,y)}{\partial(r,\theta)}\right| drd\theta = \iint f(r\cos\theta, r\sin\theta) r dr d\theta.$$

The expression in the second term is the Jacobian determinant,

$$\left|\frac{\partial(x,y)}{\partial(r,0)}\right| = \det\begin{pmatrix} \frac{\partial x}{\partial r} & \frac{\partial x}{\partial \theta} \\ \frac{\partial y}{\partial r} & \frac{\partial y}{\partial \theta} \end{pmatrix} = \det\begin{pmatrix} \cos\theta & -r\sin\theta \\ \sin\theta & r\cos\theta \end{pmatrix} = r\left(\cos^2\theta + \sin^2\theta\right) = r.$$

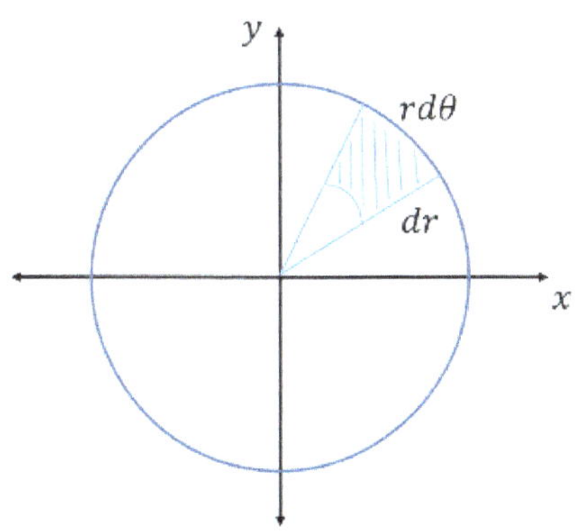

The shaded region, an area element dA , decomposed into its side lengths dr and $rd\theta$.
For infinitesimal dr and $d\theta$, $dA = rdrd\theta$ is an approximate rectangle.

Example

Integrate the function $f(r,\theta) = \frac{1}{r}$ over the annulus bounded by the circles $x^2 + y^2 = 4$ and $x^2 + y^2 = 1$.

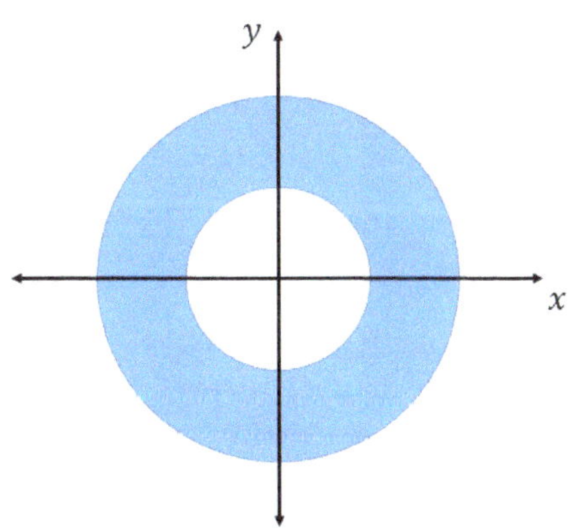

Annular region described in the problem.

In polar coordinates, the boundaries of the region are $r=1$ and $r=2$, with θ ranging from 0 to 2π over the annulus. The integral is therefore

$$\int_0^{2\pi}\int_1^2 \frac{1}{r} r dr d\theta = 2\pi\int_1^2 dr = 2\pi.$$

Triple Integrals and Higher

In higher number of dimensions, the procedure is no different; the integrals along each variable are done iteratively, and the integral limits may depend on all variables which have not yet been integrated, corresponding to some particular region of integration. In three dimensions, the triple integral $\int\int\int dxdydz$ defined by integrating 1 gives the volume of a particular region.

Example

Cube $ABCDEFGH$, labeled as shown below, has edge length 1 and is cut by a plane passing through vertex D and the midpoints M and N of $\overline{AB}$ and $\overline{CG}$ respectively. The plane divides the cube into two solids. The volume of the larger of the two solids can be written in the form $\frac{p}{q}$, where p and q are relatively prime positive integers. Find $p+q$.

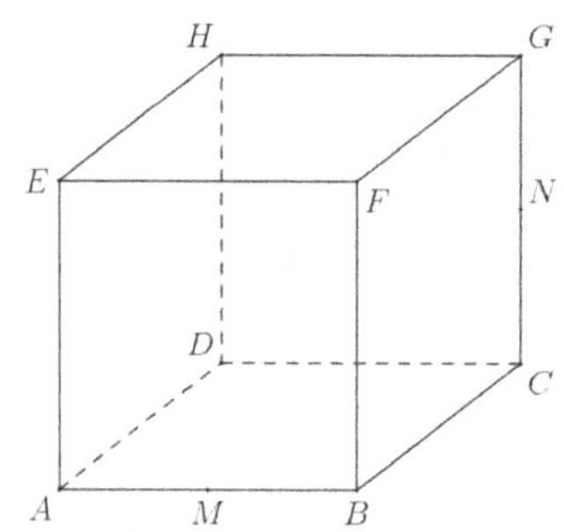

Cube referenced in the problem

Define the point M to be the origin in Cartesian coordinates. The plane is defined by the two vectors $\overline{MD}$ and $\overline{MN}$, which can be written in these coordinates

$$\overline{MD}=\left(-\frac{1}{2},1,0\right) \qquad \overline{MN}=\left(\frac{1}{2},1,\frac{1}{2}\right).$$

The normal vector to the plane is the cross product of the two

$$\overline{MD}\times\overline{MN}=\left(\frac{1}{2},\frac{1}{4},-1\right),$$

so the equation of the plane is:

$$\frac{1}{2}x+\frac{1}{4}y-z=0.$$

Looking at the *smaller* solid bounded by this plane and the cube, the z-coordinate is bounded below by 0 and above by the plane $\frac{1}{2}x+\frac{1}{4}y$. The bounds on the -coordinate are found by intersecting the plane with the $z=0$ plane: $x=-\frac{1}{2}y$ is the lower bound, while $x=1$ at the edge of the cube is the upper bound. Within these constraints there are no further constraints on y. Thus the volume of the smaller solid is

$$\int_0^1\int_{-\frac{1}{2}y}^1\int_0^{\frac{1}{2}x+\frac{1}{4}y} dz\,dx\,dy=\int_0^1\int_{-\frac{1}{2}y}^1\left(\frac{1}{2}x+\frac{1}{4}y\right)dx\,dy$$
$$=\int_0^1\left(\frac{x^2}{4}+\frac{1}{4}xy\right)\Big|_{-\frac{1}{2}y}^{1}\,dy$$
$$=\int_0^1\left(\frac{1}{4}-\frac{1}{4}y-\frac{1}{16}y^2+\frac{1}{8}y^2\right)dy$$
$$=\frac{1}{4}-\frac{1}{8}-\frac{1}{48}+\frac{1}{24}=\frac{7}{48}.$$

Subtracting from the volume of the unit cube, one finds that the volume of the larger solid is $\frac{41}{48}$, so $p+q=89$.

As in two dimensions, certain problems may be much easier using a coordinate system adapted to the problem at hand. The natural generalization of polar coordinates in 2D are cylindrical coordinates $(r,0,z)$, where r and θ are defined the same way as in 2D and the z-coordinate is left untouched.

Example

Compute the volume bounded by the cylinders $x^2+y^2=1$ and $x^2+y^2=4$ and and the planes $z=0$, $z=1$.

The relevant integral can be computed immediately:

$$\int_0^1\int_0^{2\pi}\int_1^2 r\,dr\,d\theta\,dz=3\pi.$$

Coordinates that are frequently useful in problems with spherical geometry are spherical coordinates (ρ,θ,ϕ), defined by $x=\rho\sin\phi\cos\theta$, $y=\rho\sin\phi\sin\theta$, $z=\rho\cos\phi$. Note that in physics the definitions of θ and ϕ are usually interchanged. Geometrically, ρ is the radius of a point from the origin θ, is the usual polar angle, and ϕ is the angle between a vector and the z-axis.

In spherical coordinates, the Jacobian determinant method used above shows that

$$dx\,dy\,dz=r^2\sin\phi\,d\rho\,d\theta\,d\phi$$

Towards the north pole of the sphere, pieces of volume must get smaller by the $\sin\phi$ factor because at the north pole they shrink to a single point.

Compute the integral of the function $r\sin\frac{\theta}{4}$ over the upper hemisphere of the unit sphere.

Computing from the definition,

$$\begin{aligned}\int_0^{\pi/2}\int_0^{2\pi}\int_0^1 r^3\sin\phi\sin\frac{\theta}{4}\,d\rho\,d\theta\,d\phi &= \int_0^{\pi/2}\int_0^{2\pi}\frac{1}{4}\sin\phi\sin\frac{\theta}{4}\,d\theta\,d\phi \\ &= \int_0^{\pi/2}\sin\phi\,d\phi \\ &= 1.\end{aligned}$$

Example

Evaluate the quadruple integral in four dimensions:

$$\int_0^1\int_0^1\int_0^1\int_0^1 (w+x+y+z)\,dwdxdydz.$$

The procedure for evaluating each integral iteratively is the same as previously:

$$\begin{aligned}\int_0^1\int_0^1\int_0^1\int_0^1 (w+x+y+z)\,dwdxdydz &= \int_0^1\int_0^1\int_0^1\left(\frac{1}{2}+x+y+z\right)dxdydz \\ &= \int_0^1\int_0^1\left(\frac{1}{2}+\frac{1}{2}+y+z\right)dydz \\ &= \int_0^1\left(\frac{1}{2}+\frac{1}{2}+\frac{1}{2}+z\right)dz \\ &= 2.\end{aligned}$$

Applications in Physics and Mathematics

Example

The coordinates of the center of mass of a two-dimensional object (or three-dimensional object with translation symmetry in one dimension) are given by:

$$\vec{x}\,\mathrm{COM} = \frac{1}{M}\int \sigma(\vec{x})\,\vec{x}\,dA,$$

with M the total mass of the object and $\sigma(\vec{x})$ the mass area density.

Find the (x, y) coordinates of the center of mass of the upper half of a disk of radius 1 with mass area density $\sigma = x^2 + y^2$

First, compute the total mass of the disk using polar coordinates:

$$M = \int \sigma dA = \frac{1}{2}\int_0^{2\pi}\int_0^1 r^3 dr\, d\theta = (\pi)\frac{1}{4} = \frac{\pi}{4}.$$

Now each coordinate can be computed separately. By symmetry, the x -coordinate is zero. The y -coordinate is

$$y\text{COM} = \frac{4}{\pi}\int_0^{\pi}\int_0^1 r^4 \sin\theta\, dr\, d\theta = \frac{4}{5\pi}\int_0^{\pi}\sin\theta\, d\theta = \frac{8}{5\pi}.$$

Gauss's law in electricity and magnetism reads

$$\iint \vec{E}.\ d\vec{A} = \frac{Q}{\epsilon_0},$$

where E is a vector field in three dimensions, A is some closed surface, Q is the total charge enclosed by that surface, and ϵ_0 is a constant. If Q is a spherically symmetric charge distribution, derive the electric field outside the charge distribution.

On an imaginary sphere at radius r away from the center of the charge distribution, the electric field must be the same everywhere by spherical symmetry. So the left-hand side turns into

$$\iint \vec{E}.\ d\vec{A} = E\int dA = E\left(4\pi r^2\right).$$

The surface integral turns into a double integral over the surface once the is removed; as discussed previously, this yields the area of the surface. Thus, Gauss's law says

$$E\left(4\pi r^2\right) = \frac{Q}{\epsilon_0} \Rightarrow E = \frac{Q}{4\pi\epsilon_0 r^2}.$$

This is Coulomb's law for the electric field outside a spherically symmetric ball of charge.

Example

A particle in quantum mechanics confined to two dimensions has the wavefunction:

$$\psi(x, y) = Ce^{-x^2-y^2}.$$

Find the constant C given the normalization condition $\int |\psi|^2\, dxdy = 1$.

Computing the double integral in polar coordinates,

$$\int_{-\infty}^{\infty}\int_{-\infty}^{\infty} C^2 e^{-2x^2-2y^2} dxdy = \int_0^{2\pi}\int_0^{\infty} C^2 re^{-2r^2}\, drd\theta = C^2 2\pi\left(\frac{1}{4}\right) = C^2\frac{\pi}{2}.$$

Therefore, the constant C is

$$C = \sqrt{\frac{2}{\pi}}.$$

Double Integrals Over Rectangles

Double integrals over rectangular regions are straightforward to compute in many cases. For a rectangular region S defined by xin $[a,b]$ and y in $[c,d]$, the double integral of a function $f(x,y)$ in this region is given as:

Volume to be Integrated: Double integral as volume under a surface $z = x^2 - y^2$. The rectangular region at the bottom of the body is the domain of integration, while the surface is the graph of the two-variable function to be integrated.

$$\iint_S f(x,y)dxdy = \int_a^b \left(\int_c^d f(x,y)dy \right) dx$$
$$= \int_c^d \left(\int_a^b f(x,y)dx \right) dy$$

Here, we exchanged the order of the integration, assuming that $f(x,y)$ satisfies the conditions to apply Fubini's theorem.

Example

Let us assume that we wish to integrate a multivariable function f over a region A:

$$A = \left\{ (x,y) \in \mathbf{R}^2 : 11 \le x \le 14 \ 7 \le y \le 10 \right\}$$

$$f(x,y) = x^2 + 4y$$

Formulating the double integral, we first evaluate the inner integral with respect to x:

$$\int_{11}^{14} (x^2 + 4y)\,\mathrm{d}x = (\frac{1}{3}x^3 + 4yx)\Big|_{x=11}^{x=14}\Big|$$
$$= \frac{1}{3}(14)^3 + 4y(14) - \frac{1}{3}(11)^3 - 4y(11)$$
$$= 471 + 12y$$

We then integrate the result with respect to y:

$$\int_{7}^{10} (471+12y)\,dy = (471y+6y^2)\Big|_{y=7}^{y=10}$$
$$= 471(10)+6(10)^2-471(7)-6(7)^2$$
$$= 1719$$

We could have computed the double integral starting from the integration over y. Confirm yourself that the result is the same.

Iterated Integrals

An iterated integral is the result of applying integrals to a function of more than one variable (for example $f(x,y)$ or $f(x,y,z)$ in such a way that each of the integrals considers some of the variables as given constants. For example, in the function $f(x,y)$, if y is considered a given parameter, it can be integrated with respect to x $\int f(x,y)dx$. The result is a function of y and therefore its integral can be considered again. If this is done, the result is the iterated integral:

$$\int\left(\int f(x,y)dx\right)dy$$

It is key to note that this is different, in principle, from the multiple integral $\int\int f(x,y)dxdy$. A theorem called Fubini's theorem, however, states that they may be equal under very mild conditions. The alternative notation for iterated integrals $\int dy \int f(x,y)dx$ is also used. Iterated integrals are computed following the operational order indicated by the parentheses (in the notation that uses them), starting from the innermost integral and working out.

Example

For the iterated integral $\int\left(\int f(x+y)dx\right)dy$, the integral $\int(x+y)dx = \frac{x^2}{2}+yx$ is computed first. The result is then used to compute the integral with respect to y:

$$\int\left(\frac{x^2}{2}+yx\right)dy = \frac{yx^2}{2}+\frac{xy^2}{2}$$

An iterated integral can be used to find the volume of the object in the figure

It should be noted, however, that this example omits the constants of integration. After the first integration with respect to x, we would rigorously need to introduce a "constant" function of y. That is, If we were to differentiate this function with respect to x, any terms containing only y would vanish, leaving the original integral. Similarly for the second integral, we would introduce a "constant" function of x, because we have integrated with respect to y. In this way, indefinite integration does not make much sense for functions of several variables. While the antiderivatives of single variable functions differ at most by a constant, the antiderivatives of multivariable functions differ by unknown single-variable terms, which could have a drastic effect on the behavior of the function.

Double Integrals Over General Regions

Double integrals can be evaluated over the integral domain of any general shape.

We studied how double integrals can be evaluated over a rectangular region. But there is no reason to limit the domain to a rectangular area. The integral domain can be of any general shape. In this atom, we will study how to formulate such an integral.

This method is applicable to any domain Dfor which:

- the projection of D onto either the x-axis or the y-axis is bounded by the two values, a and b.
- any line perpendicular to this axis that passes between these two values intersects the domain in an interval whose endpoints are given by the graphs of two functions, α and β.

x -axis: If the domain D is normal with respect to the x-axis, and $f: D \to R$ is a continuous function, then α(x) and β(x) (defined on the interval [a,b]) are the two functions that determine D. It follows, then, that:

$$\iint_D f(x,y)\,dxdy = \int_a^b dx \int_{\alpha(x)}^{\beta(x)} f(x,y)dy$$

y-axis: If D is normal with respect to the y-axis and $f: D \to R$ is a continuous function, then α(y) and β(y) (defined on the interval [a,b]) are the two functions that determine D. It follows, then, that

$$\iint_D f(x,y)\,dxdy = \int_a^b dy \int_{\alpha(y)}^{\beta(y)} f(x,y)dx$$

Example

Consider the following region:

$$D = \{(x,y) \in \mathbf{R}^2 \; x \geq 0, y \leq 1, y \geq x^2\}$$

Calculate $\iint_D (x+y)dxdy$. This domain is normal with respect to both the x– and y-axes. To apply the formulae, you must first find the functions that determine D and the intervals over which these are defined. In this case the two functions are $\alpha(x) = x^2$ and β(x)=1 , while the interval is given by the intersections of the functions with x=0, so the interval is $[a,b]=[0,1]$ (normality has been chosen with respect to the x-axis for a better visual understanding).

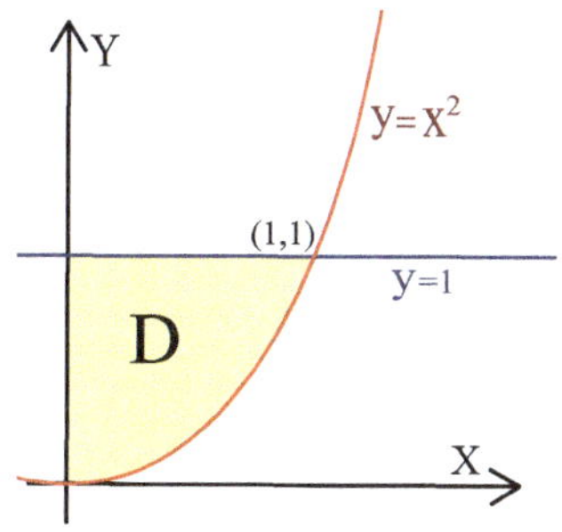

Double integral over the normal region D shown in the example

It is now possible to apply the formula:

$$\iint_D (x+y)dxdy = \int_0^1 dx \int_{x^2}^1 (x+y)dy$$
$$= \int_0^1 dx \left[xy + \frac{y^2}{2} \right]_{x^2}^1$$

(At first the second integral is calculated considering xx as a constant). The remaining operations consist of applying the basic techniques of integration:

$$\int_0^1 \left[xy + \frac{y^2}{2} \right]_{x^2}^1 dx = \int_0^1 \left(x + \frac{1}{2} - x^3 - \frac{x^4}{2} \right) = \frac{13}{20}$$

Double Integrals in Polar Coordinates

When domain has a cylindrical symmetry and the function has several specific characteristics, apply the transformation to polar coordinates.

In R^2, if the domain has a cylindrical symmetry and the function has several particular characteristics, you can apply the transformation to polar coordinates, which means that the generic points P(x,y)in Cartesian coordinates switch to their respective points in polar coordinates. This allows one to change the shape of the domain and simplify the operations.

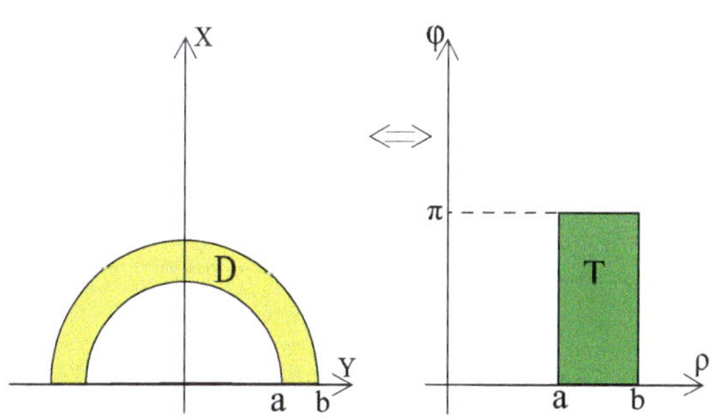

This figure illustrates graphically a transformation from cartesian to polar coordinates

Change of Variable

The polar coordinates r and φ can be converted to the Cartesian coordinates x and y by using the trigonometric functions sine and cosine:

$$x = r\cos\varphi$$
$$y = r\sin\varphi$$

The Cartesian coordinates x and y can be converted to polar coordinates r and φ with $r \geq 0$ and φ in the interval $(-\pi, \pi]$:

$$r = \sqrt{x^2 + y^2}$$
$$\varphi = \tan^{-1}(\frac{y}{x})$$

The fundamental relation to make the transformation is as follows:

$$f(x, y) \rightarrow f(\rho\cos\phi, \rho\sin\phi)$$

Examples

- Given the function $f(x, y) = x + y$ and applying the transformation, one obtains $f(\rho,\phi) = \rho\cos\phi + \rho\sin\phi = \rho(\cos\phi + \sin\phi)$.
- Given the function $f(x, y) = x^2 + y^2$, one can obtain $f(\rho,\phi) = \rho^2(\cos^2\phi + \sin^2\phi) = \rho^2$ using the Pythagorean trigonometric identity, which is very useful to simplify this operation. Particularly in this case, you can see that the representation of the function f became simpler in polar coordinates. This is the case because the function has a cylindrical symmetry. In general, the best practice is to use the coordinates that match the built-in symmetry of the function.

Integrals in Polar Coordinates

The Jacobian determinant of that transformation is the following:

$$\frac{\partial(x, y)}{\partial(\rho,\phi)} = \begin{vmatrix} \cos\phi & -\rho\sin\phi \\ \sin\phi & \rho\cos\phi \end{vmatrix} = \rho$$

which has been obtained by inserting the partial derivatives of $x = \rho\cos(\varphi)$, $y = \rho sin(\varphi)$ in the first column with respect to ρ and in the second column with respect to φ, so the dxdy differentials in this transformation become $\rho d\rho d\varphi$. Once the function is transformed and the domain evaluated, it is possible to define the formula for the change of variables in polar coordinates:

$$\iint_D f(x, y)dxdy = \iint_T f(\rho\cos\varphi, \rho\sin\varphi)\rho$$

Example

Integrate the function $f(x, y) = x$ over the domain:

$$D = \{x^2 + y^2 \leq 9, x^2 + y^2 \geq 4, y \geq 0\}$$

From $f(x,y) = x \rightarrow f(\rho,\phi) = \rho\cos\phi$

$$\iint_D x\,dx\,dy = \iint_T \rho\cos\phi\rho d\rho d\phi$$

$$= \int_0^{\pi}\int_2^3 \rho^2\cos\phi d\rho d\phi$$

$$= \int_0^{\pi}\cos\phi d\phi\left[\frac{\rho^3}{3}\right]_2^3$$

$$= [\sin\phi]_0^{\pi}\left(9 - \frac{8}{3}\right) = 0$$

Triple Integrals in Cylindrical Coordinates

When the function to be integrated has a cylindrical symmetry, it is sensible to change the variables into cylindrical coordinates and then perform integration.

In R^3 the integration on domains with a circular base can be made by the passage in cylindrical coordinates; the transformation of the function is made by the following relation:

$$f(x,y,z) \rightarrow f(\rho\cos\varphi, \rho\sin\varphi, z)$$

The domain transformation can be graphically attained, because only the shape of the base varies, while the height follows the shape of the starting region. Also in switching to cylindrical coordinates, the $dxdydz$ differentials in the integral become $\rho d\rho d\varphi dz$.

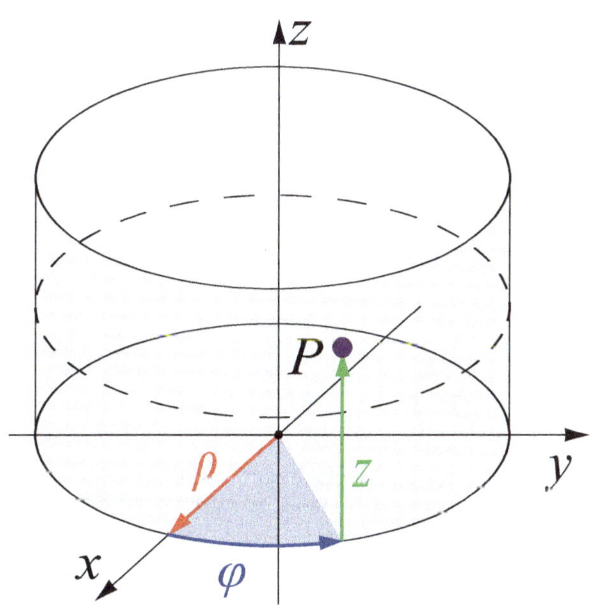

Cylindrical coordinates are often used for integrations on domains with a circular base.

Example

The region is:

$$D = \{x^2 + y^2 \leq 9, x^2 + y^2 \geq 4, 0 \leq z \leq 5\}$$

If the transformation is applied, this region is obtained:

$$T = \{2 \leq \rho \leq 3, 0 \leq \varphi \leq 2\pi, 0 \leq z \leq 5\}$$

because the z component is unvaried during the transformation, the $dxdydz$ differentials vary as in the passage in polar coordinates: therefore, they become: $\rho d\rho d\varphi dz$ c. Finally, it is possible to apply the final formula to cylindrical coordinates:

$$\iiint Df(x,y,z)dxdydz = \iiint Tf(\rho\cos\varphi, \rho\sin\varphi, z)\rho d\rho d\varphi dz$$

This method is convenient in case of cylindrical or conical domains or in regions where it is easy to individuate the z interval and even transform the circular base and the function.

Example

The function $f(x,y,z) = x2 + y2 + z$ is and as integration domain this cylinder:

$$D = \{x^2 + y^2 \leq 9, -5 \leq z \leq 5\}$$

The transformation of D in cylindrical coordinates is the following:

$$T = \{0 \leq \rho \leq 3, 0 \leq \phi \leq 2\pi, -5 \leq z \leq 5\}$$

while the function becomes:

$$f(\rho\cos\varphi, \rho\sin\varphi, z) = \rho^2 + z$$

Therefore, the integral becomes:

$$\iiint_D (x^2 + y^2 + z)dx\,dy\,dz = \iiint_T (\rho^2 + z)\rho d\rho d\phi dz$$

$$= \int_{-5}^{5} dz \int_0^{2\pi} d\phi \int_0^3 (\rho 3 + \rho z)\, d\rho$$

$$= 2\pi \int_{-5}^{5} \left[\frac{\rho^4}{4} + \frac{\rho^2 z}{2}\right]_0^3 dz$$

$$= 2\pi \int_{-5}^{5} \left(\frac{81}{4} + \frac{9}{2} z\right) dz$$

$$= 405\,\pi$$

Triple Integrals in Spherical Coordinates

When the function to be integrated has a spherical symmetry, it is sensible to change the variables into spherical coordinates and then perform integration.

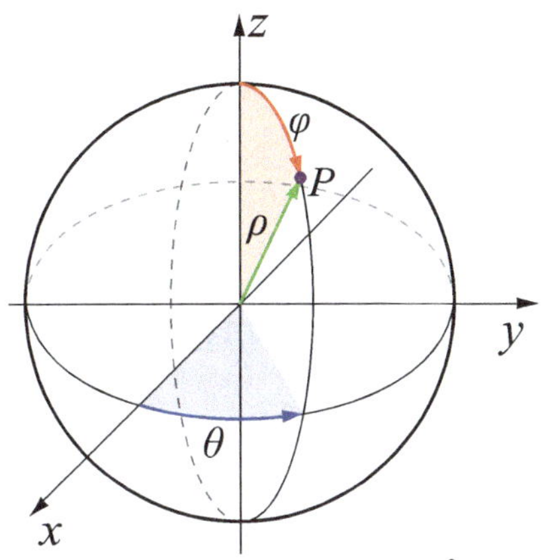

Spherical coordinates are useful when domains in R^3 have spherical symmetry.

In R^3 some domains have a spherical symmetry, so it's possible to specify the coordinates of every point of the integration region by two angles and one distance. It's possible to use therefore the passage in spherical coordinates; the function is transformed by this relation:

$$f(x,y,z) \rightarrow f(\rho\cos\theta\sin\varphi, \rho\sin\theta\sin\varphi, \rho\cos\varphi)$$

Points on z-axis do not have a precise characterization in spherical coordinates, so θ can vary from 0 to 2π. The better integration domain for this passage is obviously the sphere.

Integrals in Spherical Coordinates

The Jacobian determinant of this transformation is the following:

$$\frac{\partial(x,y,z)}{\partial(\rho,\theta,\varphi)} = \begin{vmatrix} \cos\theta\sin\varphi & -\rho\sin\theta\sin\varphi & \rho\cos\theta\cos\varphi \\ \sin\theta\sin\varphi & \rho\cos\theta\sin\varphi & \rho\sin\theta\cos\varphi \\ \cos\varphi & 0 & -\rho\sin\varphi \end{vmatrix} = \rho^2\sin\varphi$$

The $dxdydz$ differentials therefore are transformed to $\rho^2\sin\varphi d\rho d\varphi dz$.

Finally, you obtain the final integration formula: It's better to use this method in case of spherical domains and in case of functions that can be easily simplified, by the first fundamental relation of trigonometry, extended in R^3 ; in other cases it can be better to use cylindrical coordinates.

Example

Integrate $f(x,y,z) = x2 + y2 + z2$ over the domain $D = x^2 + y^2 + z^2 \le 16$.

In spherical coordinates:

$$f(\rho\sin\varphi\cos\theta, \rho\sin\varphi\sin\theta, \rho\cos\varphi) = \rho^2$$

while the intervals of the transformed region T from D :

$$0 \le \rho \le 4, 0 \le \varphi \le \pi, 0 \le \theta \le 2\pi$$

Therefore:

$$\iiint_D (x^2 + y^2 + z^2)\, dx\, dy\, dz = \iiint_T \rho^2\, \rho^2 \sin\theta\, d\rho\, d\theta\, d\phi,$$

$$= \int_0^{\pi} \sin\phi\, d\phi \int_0^4 \rho^4\, d\rho \int_0^{2\pi} d\theta$$

$$= 2\pi \int_0^{\pi} \sin\phi \left[\frac{\rho^5}{5}\right]_0^4 d\phi$$

$$= 2\pi \left[\frac{\rho^5}{5}\right]_0^4 \left[-\cos\phi\right]_0^{\pi} = \frac{4096\pi}{5}$$

Triple Integrals

For $T \subseteq R^3$, the triple integral over T is written as

$$\iiint_T f(x, y, z)\, dx dy dz$$

Notice that, by convention, the triple integral has three integral signs (and a double integral has two integral signs); this is a notational convention which is convenient when computing a multiple integral as an iterated integral.

We have seen that double integrals can be evaluated over regions with a general shape. The extension of those formulae to triple integrals should be apparent. If T is a domain that is normal with respect to the *xy*-plane and determined by the functions $\alpha(x, y)$ and $\beta(x, y)$, then:

$$\iiint_T f(x, y, z)\, dx\, dy\, dz = \iint_D \int_{\alpha(x,y)}^{\beta(x,y)} f(x, y, z)\, dz dx dy.$$

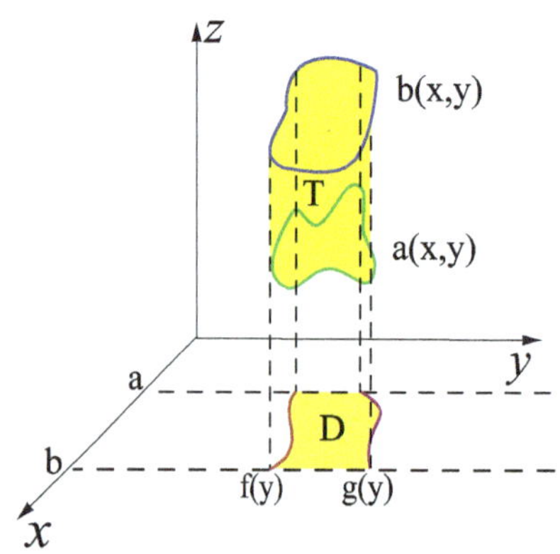

Graphical Representation of a Triple Integral:
Example of domain in R3 that is normal with respect to the xy-plane.

Example

The volume of the parallelepiped of sides 4 by 6 by 5 may be obtained in two ways:

- By calculating the double integral of the function f(x,y)=5 over the region D in the xy-plane which is the base of the parallelepiped: $\iint_D 5\,dxdy$
- By calculating the triple integral of the constant function 1 over the parallelepiped itself: $\iiint_{\text{parallelepiped}} 1dxdydz$

Example

Integrate $f(x,y,z)=x^2+y^2+z^2$ over the domain $D=\{x^2+y^2+z^2\leq 16\}$.

Looking at the domain, it seems convenient to adopt the passage in spherical coordinates; in fact, the intervals of the variables that delimit the new T region are obviously:

$$(0\leq\rho\leq 4, 0\leq\phi\leq\pi, 0\leq\theta\leq 2\pi)$$

For the function, we get:

$$f(\rho\sin\phi\cos\theta, \rho\sin\phi\sin\theta, \rho\cos\phi)=\rho^2$$

Therefore:

$$\iiint_D (x^2+y^2+z^2)dxdydz = \iiint_T \rho^2\,\rho^2\sin\theta\,d\rho\,d\theta\,d\phi$$

$$=\int_0^\pi \sin\phi d\phi \int_0^4 \rho^4 d\rho \int_0^{2\pi} d\theta$$

$$=2\pi\int_0^\pi \sin\phi\left[\frac{\rho^5}{5}\right]_0^4 d\phi$$

$$=2\pi\left[\frac{\rho^5}{5}\right]_0^4 [-\cos\phi]_0^\pi = \frac{4096\pi}{5}$$

Change of Variables

One makes a change of variables to rewrite the integral in a more "comfortable" region, which can be described in simpler formulae

The limits of integration are often not easily interchangeable (without normality or with complex formulae to integrate). One makes a change of variables to rewrite the integral in a more "comfortable" region, which can be described in simpler formulae. To do so, the function must be adapted to the new coordinates.

For example, for the function $f(x,y)=(x-1)^2+\sqrt{y}$, if one adopts this substitution $x'=x-1, y'=y$, therefore $=x'+1, y=y'$, one obtains the new function:

$$f_2(x,y)=(x')^2+\sqrt{y}$$

which is simpler than the original form. When changing integration variables, however, make sure that the integral domain also changes accordingly. In the example, if integration is performed over x in [0,1] and y in [0,3], the new variables x′and y′ vary over [−1,0] and [0,3], respectively. There exist three main "kinds" of changes of variable (one in R^2 , two in R^3); however, more general substitutions can be made using the same principle.

1. Polar Coordinates

The function to be integrated transforms as:

$$f(x,y) \to f(\rho\cos\phi, \rho\sin\phi)$$

and the integral accordingly changes as:

$$\iint_D f(x,y)\,dx\,dy = \iint_T f(\rho\cos\phi, \rho\sin\phi)\rho\,d\rho\,d\phi$$

2. Cylindrical Coordinates

The function to be integrated transforms as:

$$f(x,y,z) \to f(\rho\cos\phi, \rho\sin\phi, z)$$

and the integral accordingly changes as:

$$\iiint_D f(x,y,z)\,dx\,dy\,dz \quad \iiint_T f(\rho\cos\phi, \rho\sin\phi, z)\rho\,d\rho\,d\phi\,dz$$

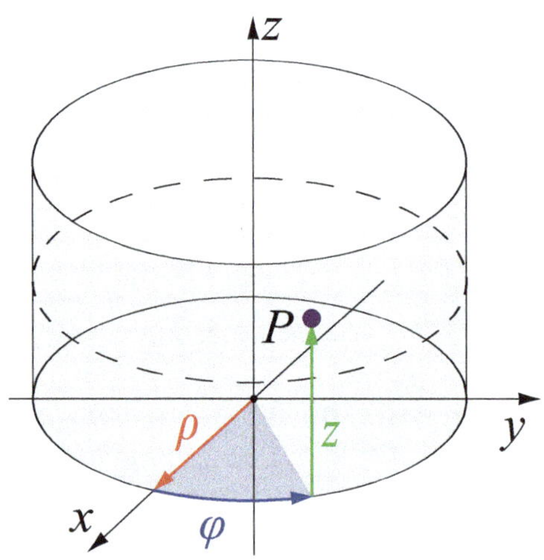

Changing to cylindrical coordinates may be useful depending on the setup of problem.

3. Spherical coordinates

The function to be integrated transforms as:

$$f(x,y,z) \to f(\rho\cos\theta\sin\phi, \rho\sin\theta\sin\phi, \rho\cos\phi)$$

and the integral accordingly changes as:

$$\iiint_D f(x,y,z)dxdydz$$
$$= \iiint_T f(\rho sin\phi cos\theta, \rho sin\phi sin\theta, \rho cos\phi)\rho^2 sin\phi d\rho d\theta d\phi$$

Multiple Integrals and Iterated Integrals

Fubini's theorem states that if

$$\iint_{A\times B} |f(x,y)| d(x,y) < \infty,$$

that is, if the integral is absolutely convergent, then the multiple integral will give the same result as the iterated integral,

$$\iint_{A\times B} f(x,y)d(x,y) = \int_A \left(\int_B f(x,y)dy\right)dx = \int_B \left(\int_A f(x,y)dx\right)dy.$$

In particular this will occur if $|f(x, y)|$ is a bounded function and A and B are bounded sets.

If the integral is not absolutely convergent, care is needed not to confuse the concepts of *multiple integral* and *iterated integral*, especially since the same notation is often used for either concept. The notation

$$\int_0^1 \int_0^1 f(x,y)dydx$$

means, in some cases, an iterated integral rather than a true double integral. In an iterated integral, the outer integral

$$\int_0^1 \cdots dx$$

is the integral with respect to x of the following function of x:

$$g(x) = \int_0^1 f(x,y)dy.$$

A double integral, on the other hand, is defined with respect to area in the xy-plane. If the double integral exists, then it is equal to each of the two iterated integrals (either "$dy\ dx$" or "$dx\ dy$") and one often computes it by computing either of the iterated integrals. But sometimes the two iterated integrals exist when the double integral does not, and in some such cases the two iterated integrals are different numbers, i.e., one has

$$\int_0^1 \int_0^1 f(x,y)dydx \neq \int_0^1 \int_0^1 f(x,y)dxdy.$$

This is an instance of rearrangement of a conditionally convergent integral.

On the other hand, some conditions ensure that the two iterated integrals are equal even though the double integral need not exist. By the Fichtenholz–Lichtenstein theorem, if *f* is bounded on [0, 1] × [0, 1] and both iterated integrals exist, then they are equal. Moreover, existence of the inner integrals ensures existence of the outer integrals. The double integral need not exist in this case even as Lebesgue integral, according to Sierpiński.

The notation

$$\int_{[0,1]\times[0,1]} f(x,y)dxdy$$

may be used if one wishes to be emphatic about intending a double integral rather than an iterated integral.

Some Practical Applications

Quite generally, just as in one variable, one can use the multiple integral to find the average of a function over a given set. Given a set $D \subseteq \mathbf{R}^n$ and an integrable function f over D, the average value of f over its domain is given by

$$\bar{f} = \frac{1}{m(D)} \int_D f(x)dx,$$

where $m(D)$ is the measure of D.

Additionally, multiple integrals are used in many applications in physics. The examples below also show some variations in the notation.

In mechanics, the moment of inertia is calculated as the volume integral (triple integral) of the density weighed with the square of the distance from the axis:

$$I_z = \iiint_V \rho r^2 \, dV.$$

The gravitational potential associated with a mass distribution given by a mass measure dm on three-dimensional Euclidean space R^3 is

$$V(\mathbf{x}) = -\iiint_{\mathbf{R}^3} \frac{G}{|\mathbf{x}-\mathbf{y}|} dm(\mathbf{y}).$$

If there is a continuous function $\rho(\mathbf{x})$ representing the density of the distribution at $\mathbf{x}$, so that $dm(\mathbf{x}) = \rho(\mathbf{x})d^3\mathbf{x}$, where $d^3\mathbf{x}$ is the Euclidean volume element, then the gravitational potential is

$$V(\mathbf{x}) = -\iiint_{\mathbf{R}^3} \frac{G}{|\mathbf{x}-\mathbf{y}|} \rho(\mathbf{y}) d^3\mathbf{y}.$$

In electromagnetism, Maxwell's equations can be written using multiple integrals to calculate the total magnetic and electric fields. In the following example, the electric field produced by a distribution of charges given by the volume charge density $\rho(\vec{r})$ is obtained by a *triple integral* of a vector function:

$$\vec{E} = \frac{1}{4\pi\varepsilon_0} \iiint \frac{\vec{r} - \vec{r}'}{\|\vec{r} - \vec{r}'\|^3} \rho(\vec{r}') d^3 r'.$$

This can also be written as an integral with respect to a signed measure representing the charge distribution.

Multiple integrals are used in many applications in physics and engineering.

As is the case with one variable, one can use the multiple integral to find the average of a function over a given set. Given a set $D \subseteq R^n$ and an integrable function f over D ,the average value of f over its domain is given by:

$$\bar{f} = \frac{1}{m(D)} \int_D f(x)\, dx$$

where $m(D)$ is the measure of D Additionally, multiple integrals are used in many applications in physics and engineering. The examples below also show some variations in the notation.

Example

In mechanics, the moment of inertia is calculated as the volume integral (triple integral) of the density weighed with the square of the distance from the axis:

$$I_z = \iiint_V \rho r^2 dV$$

Example

The gravitational potential associated with a mass distribution given by a mass measure *dm* on three-dimensional Euclidean space $R^3$is:

$$V(x) = -\int_{R^3} \frac{G}{|x - r|} dm(r)$$

If there is a continuous function $\rho(x)$ representing the density of the distribution at x, so that dm $dm(x) = \rho(x) d^3 x$, where $d^3 x$ is the Euclidean volume element, then the gravitational potential is:

$$V(x) = -\int_{R^3} \frac{G}{|x - r|} \rho(r) d^3 r$$

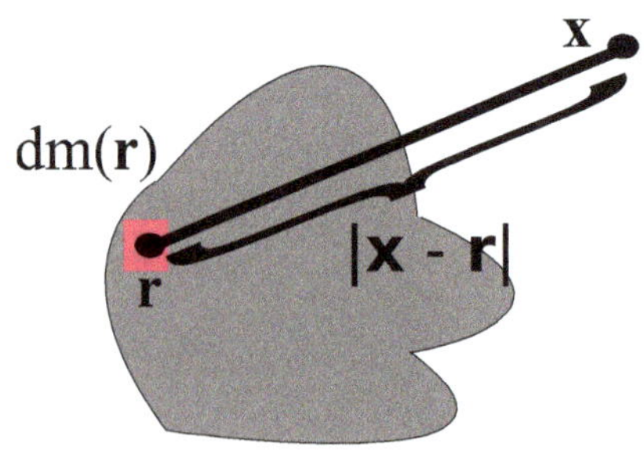

A Mass to be Integrated: Points x and r, with r contained in the distributed mass (gray) and differential mass dm(r) located at the point r.

Example

In electromagnetism, Maxwell's equations can be written using multiple integrals to calculate the total magnetic and electric fields. In the following example, the electric field produced by a distribution of charges given by the volume charge density $\rho(\vec{r}')$ is obtained by a triple integral of a vector function:

$$\vec{E} = \frac{1}{4\pi \in_0} \iiint \frac{\vec{r} - \vec{r}'}{\|\vec{r} - \vec{r}'\|^3} \rho(\vec{r}')\, d^3 r'$$

This can also be written as an integral with respect to a signed measure representing the charge distribution.

Center of Mass and Inertia

The center of mass for a rigid body can be expressed as a triple integral.

Multiple integrals are used in many applications in physics and engineering. In this atom, we will see how center of mass can be calculated using multiple integrals.

The center of mass is the unique point at the center of a distribution of mass in space that has the property that the weighted position vectors relative to this point sum to zero. In the case of a single rigid body, the center of mass is fixed in relation to the body, and if the body has uniform density, it will be located at the centroid. The center of mass may be located outside the physical body, as is sometimes the case for hollow or open-shaped objects, such as a horseshoe. In the case of a distribution of separate bodies, such as the planets of the Solar System, the center of mass may not correspond to the position of any individual member of the system.

A System of Particles

In the case of a system of particles $P_i, i = 1, \ldots, n$, each with mass m_i that are located in space with coordinates $\mathbf{r}_i, i = 1, \ldots, n$, the coordinates $\mathbf{R}$ of the center of mass satisfy the condition:

$$\sum_{i=1}^{n} m_i (\mathbf{r}_i - \mathbf{R}) = 0$$

Solve this equation for $\mathbf{R}$ to obtain the formula:

$$\mathbf{R} = \frac{1}{M}\sum_{i=1}^{n} m_i \mathbf{r}_i$$

where M is the sum of the masses of all of the particles.

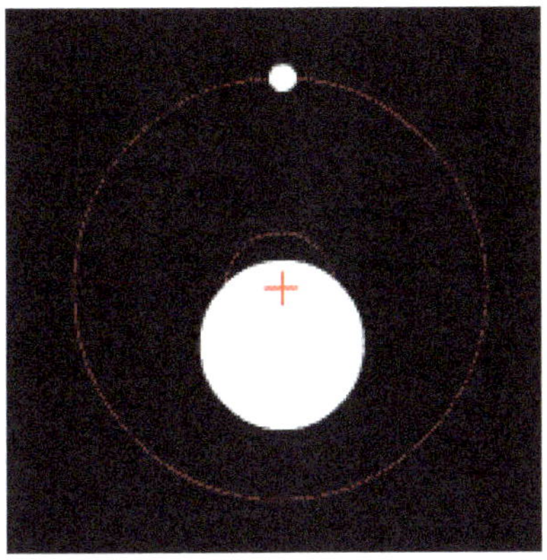

Center of Mass: Two bodies orbiting around the center of mass inside one body

A Continuous Volume

If the mass distribution is continuous with the density ρ(r) within a volume *V*, then the integral of the weighted position coordinates of the points in this volume relative to the center of mass **R** is zero; that is:

$$\int_V \rho(r)(\mathbf{r} - \mathbf{R})dV = 0$$

Solve this equation for the coordinates **R** to obtain:

$$\mathbf{R} = \frac{1}{M}\int_V \rho(r)\,\mathbf{r}dV$$

where M is the total mass in the volume. The integral is over the three dimensional volume, so it is a triple integral.

Surface Integral

Similar to the integral of a scalar field over a curve, which we called the line integral, we can define the integral of a vector-field over a surface.

Let S be a surface in space with finite surface area. Let f be a continuous scalar-field defined on the surface S .We can subdivide S into smaller portions, say $S_1, S_2, \ldots, S_n,$ having areas , and form the sum

$$\sigma_k := \sum_{K=1}^{n} f(x_k, y_k, z_k)\Delta S_{k,}$$

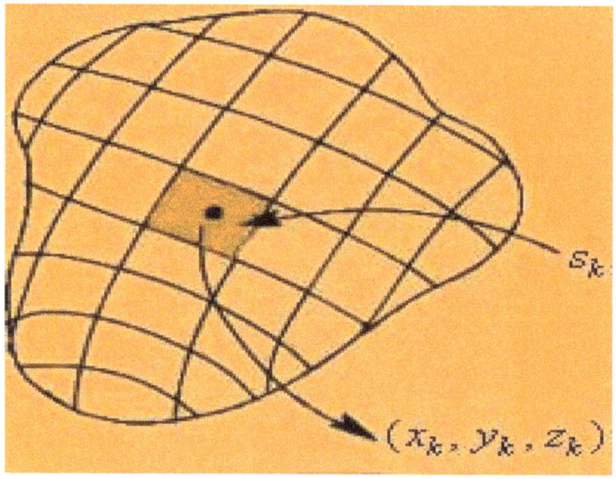

Subdivision of the surface

where $(x_k, y_k, z_k) \in S_{k,}$ is selected arbitrarily. By refining the patches into more smaller patches such that $\max(\Delta S_{k,}) \to 0$, if σ_k approaches a limit, we call it the surface integral of f over S, and denote it by

$$\iint_S f(x, y, z)\, dS.$$

Let S be a surface with parameterization

$$\mathbf{r} : R \subseteq \mathbb{R}^2 \to \mathbb{R}^2, (u, v) \mapsto \mathbf{r}(u, v), (u, v) \in R.$$

If $\mathbf{r}(u, v)$ is continuous and R is closed and bounded, then for a continuous function $f : R \subseteq \mathbb{R}^3$, we can define

$$\iint_S f(x, y, z)\, dS := \iint_R f(x(u, v), y(u, v), z(u, v)) \|\mathbf{r}_u \times \mathbf{r}_v\|\, du\, dv,$$

called the surface integral of f over the surface S

Example:

Let us evaluate the surface integral

$$\iint_S y^2\, dS,$$

where S is the sphere

$$\{(x, y, z) \in \mathbb{R}^3 \,|\, x^2 + y^2 + z^2 = 1\}$$

We give S the spherical coordinate parameterization

$$\mathbf{r}(\theta, \phi) = \cos\theta \sin\phi\, \mathbf{i} + \sin\theta \sin\phi + \cos\phi \mathbf{k}, (\theta, \phi) \in [0, 2\pi] \times [0, \pi]$$

Then

$$\mathbf{r}_\theta = -\sin\theta\sin\phi\,\mathbf{i} + \cos\theta\sin\phi\,\mathbf{j},$$

And

$$\mathbf{r}_\phi = -\cos\theta\cos\phi\,\mathbf{i} + \sin\theta\cos\phi\,\mathbf{j} - \sin\phi\mathbf{k}.$$

Thus

$$\mathbf{r}_\theta \times \mathbf{r}_\phi = \begin{vmatrix} \mathbf{i} & \mathbf{j} & \mathbf{k} \\ -\sin\theta\sin\phi & \cos\theta\sin\phi & 0 \\ \cos\theta\cos\phi & \sin\theta\cos\phi & -\sin\phi \end{vmatrix}$$

$$= \left(-\sin^2\phi\cos\theta\right)\mathbf{i} - \left(+\sin^2\phi\sin\theta\right)\mathbf{j}$$

$$+ \left(-\sin^2\theta\sin\phi\cos\phi - \cos^2\theta\sin\ \phi\cos\theta\right)\mathbf{k}.$$

Hence,

$$\left\|\mathbf{r}_\theta \times \mathbf{r}_\phi\right\|^2 = \sin^4\phi\left(\cos^2\theta\sin^2\theta\right) + \sin^2\phi\cos^3\phi$$

$$= \sin^2\phi\left(\sin^2\phi + \cos^2\phi\right) = \sin^2\phi.$$

Thus,

$$\left\|\mathbf{r}_\theta \times \mathbf{r}_\phi\right\| = \sin\phi$$

This gives, for $R = [0, 2\pi] \times [0, \pi]$,

$$\iint_S y^2 ds = \iint_R \sin^2\phi\sin^2\theta\sin\phi\, d\theta d\phi$$

$$= \int_0^{2\pi}\int_0^{\pi} \sin^3\phi\sin^2\theta\, d\theta d\phi$$

$$= \int_0^{2\pi}\left[\int_0^{\pi} \sin^3\, d\phi\right]\sin^2\theta\, d\theta$$

$$= \int_0^{2\pi}\left[\frac{1}{3}\cos^3\phi - \cos\ \phi\right]_0^{\pi} \sin^2\theta\, d\theta$$

$$= \frac{4}{3}\int_0^{2\pi} \sin^2\theta\, d\theta$$

$$= \frac{4}{3}\left[\frac{0}{2} + \sin 2\theta\right]_0^{2\pi}$$

$$= \frac{4\pi}{3}.$$

Surface Integral for Surfaces in Explicit form

For a smooth surface given explicitly as

$$S = \{(x, y, z) | z = h(x, y) \text{ for } (x, y) \in D\},$$

a parameterization is given by

$$\mathbf{r}(x, y) = x\,\mathbf{i} + y\,\mathbf{j} + h(x, y)\mathbf{k}, (x, y) \in D.$$

Since,

$$\|\mathbf{r}_x \times \mathbf{r}_y\| = \sqrt{1 + h_x^2 + h_y^2}$$

we have

$$\iint_S f(x, y, z)\,dS = \iint_R f(x, y, h(x, y))\sqrt{1 + h_x^2 + h_y^2}\;dxdy.$$

Similarly, if S is given by

$$S = \{(x, y, z) | x = g(y, z), (y, z) \in R\},$$

Then

$$\iint_S f(x, y, z)\,dS = \iint_R f(g(y, z), y, z)\sqrt{1 + g_y^2 + g_z^2}\;dy\,dz.$$

Finally, if S is given by

$$S = \{(x, y, z) | y = k(x, y), (x, y) \in R\},$$

Then

$$\iint_S f(x, y, z)\,dS = \iint_R f(x, k(x, z), z)\sqrt{1 + k_x^2 + k_z^2}\;dy\,dz.$$

Example :

Let us evaluate the integral

$$\iint_R f dS,$$

where $f(x, y, z) = z^2$ and S isthesurfaceofthecone $z^2 = x^2 + y^2$ between the planes $z = 1$ and $z = 2$. We can give the surface the following parameterization

$$\mathbf{r}(x, y) = x\,\mathbf{i} + y\,\mathbf{j} + \left(\sqrt{x^2, y^2}\right)\mathbf{k}, (x, y) \in R,$$

where R is the projection of the surface on the xy -plane

$$R := \{(x, y) | 1 \leq x^2 + y^2 \leq 2\}.$$

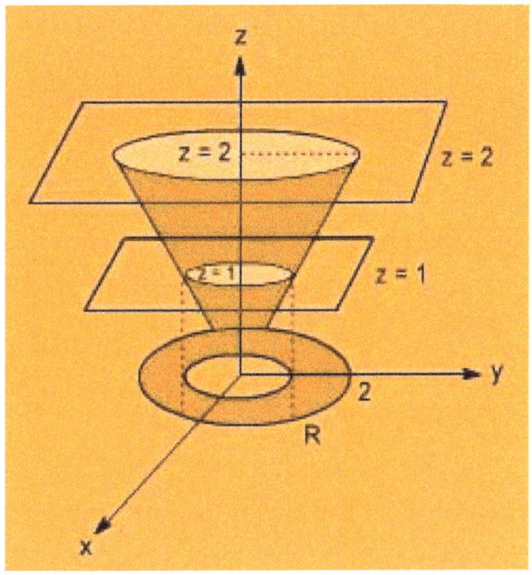

Figure: Surface S and its projection R ,

Since $\|\mathbf{r}_x \times \mathbf{r}_y\| = \sqrt{2},$ we have

$$\iint_S f\,ds = \iint_R (x^2 + y^2)\sqrt{2}\,dx\,dy$$

$$= \int_{r=1}^{2}\int_0^{2\pi} \sqrt{2}\,r^2 r\,dr d\theta \text{(Using polar coordinates)}$$

$$= 2\sqrt{2}\,\pi \int_1^2 r^3 dr$$

$$= 2\sqrt{2}\,\pi \left[\frac{16}{4} - \frac{1}{4}\right]$$

$$= \frac{15\sqrt{2}\,\pi}{2}$$

Surface Integrals of Scalar Fields

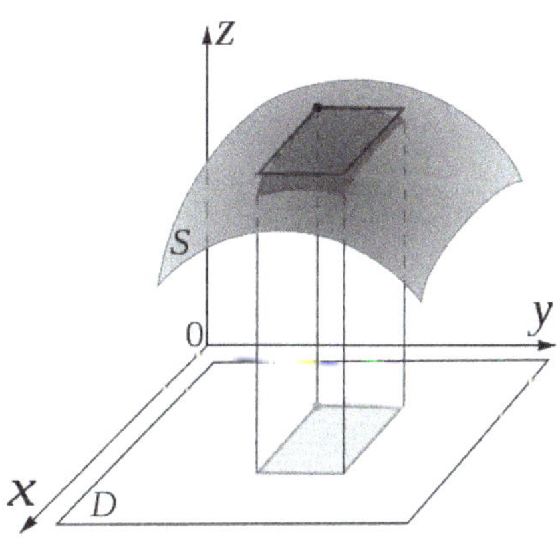

An illustration of a single surface element. These elements are made infinitesimally small, by the limiting process, so as to approximate the surface.

To find an explicit formula for the surface integral, we need to parameterize the surface of interest, *S*, by considering a system of curvilinear coordinates on *S*, like the latitude and longitude on a sphere. Let such a parameterization be x(*s*, *t*), where (*s*, *t*) varies in some region *T* in the plane. Then, the surface integral is given by

$$\iint_S f \, d\Sigma = \iint_T f(\mathbf{x}(s,t)) \left\| \frac{\partial \mathbf{x}}{\partial s} \times \frac{\partial \mathbf{x}}{\partial t} \right\| ds\, dt$$

where the expression between bars on the right-hand side is the magnitude of the cross product of the partial derivatives of **x**(*s*, *t*), and is known as the surface element. The surface integral can also be expressed in the equivalent form

$$\iint_S f \, d\Sigma = \iint_T f(\mathbf{x}(s,t)) \sqrt{g}\, ds\, dt$$

where *g* is the determinant of the first fundamental form of the surface mapping x(*s*, *t*).

For example, if we want to find the surface area of the graph of some scalar function, say $z = f(x, y)$, we have

$$A = \iint_S d\Sigma = \iint_T \left\| \frac{\partial \mathbf{r}}{\partial x} \times \frac{\partial \mathbf{r}}{\partial y} \right\| dx\, dy$$

where $\mathbf{r} = (x, y, z) = (x, y, f(x, y))$. So that $\frac{\partial \mathbf{r}}{\partial x} = (1, 0, f_x(x, y))$, and $\frac{\partial \mathbf{r}}{\partial y} = (0, 1, f_y(x, y))$. So,

$$\begin{aligned} A &= \iint_T \left\| \left(1, 0, \frac{\partial f}{\partial x}\right) \times \left(0, 1, \frac{\partial f}{\partial y}\right) \right\| dx\, dy \\ &= \iint_T \left\| \left(-\frac{\partial f}{\partial x}, -\frac{\partial f}{\partial y}, 1\right) \right\| dx\, dy \\ &= \iint_T \sqrt{\left(\frac{\partial f}{\partial x}\right)^2 + \left(\frac{\partial f}{\partial y}\right)^2 + 1}\, dx\, dy \end{aligned}$$

which is the standard formula for the area of a surface described this way. One can recognize the vector in the second line above as the normal vector to the surface.

Because of the presence of the cross product, the above formulas only work for surfaces embedded in three-dimensional space.

This can be seen as integrating a Riemannian volume form on the parameterized surface, where the metric tensor is given by the first fundamental form of the surface.

Surface Integrals of Vector Fields

Consider a vector field v on *S*, that is, for each x in *S*, v(x) is a vector.

The surface integral can be defined component-wise according to the definition of the surface integral of a scalar field; the result is a vector. This applies for example in the expression of the electric field at some fixed point due to an electrically charged surface, or the gravity at some fixed point due to a sheet of material.

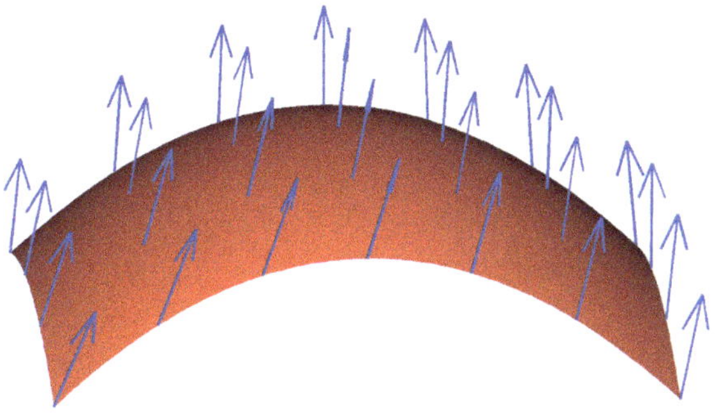

A vector field on a surface

Alternatively, if we integrate the normal component of the vector field, the result is a scalar. Imagine that we have a fluid flowing through S, such that $\mathbf{v}(\mathbf{x})$ determines the velocity of the fluid at $\mathbf{x}$. The flux is defined as the quantity of fluid flowing through S per unit time.

This illustration implies that if the vector field is tangent to S at each point, then the flux is zero because the fluid just flows in parallel to S, and neither in nor out. This also implies that if $\mathbf{v}$ does not just flow along S, that is, if $\mathbf{v}$ has both a tangential and a normal component, then only the normal component contributes to the flux. Based on this reasoning, to find the flux, we need to take the dot product of $\mathbf{v}$ with the unit surface normal $\mathbf{n}$ to S at each point, which will give us a scalar field, and integrate the obtained field as above. We find the formula

$$\begin{aligned}\iint_S \mathrm{v}\cdot \mathrm{d}\Sigma &= \iint_S (\mathrm{v}\cdot \mathrm{n})\mathrm{d}\Sigma \\ &= \iint_T \left(\mathbf{v}(\mathbf{x}(s,t)) \cdot \frac{\left(\frac{\partial \mathbf{x}}{\partial s}\times\frac{\partial \mathbf{x}}{\partial t}\right)}{\left\|\left(\frac{\partial \mathbf{x}}{\partial s}\times\frac{\partial \mathbf{x}}{\partial t}\right)\right\|}\right) \left\|\left(\frac{\partial \mathbf{x}}{\partial s}\times\frac{\partial \mathbf{x}}{\partial t}\right)\right\| \mathrm{d}s\,\mathrm{d}t \\ &= \iint_T \mathbf{v}(\mathbf{x}(s,t)) \cdot \left(\frac{\partial \mathbf{x}}{\partial s}\times\frac{\partial \mathbf{x}}{\partial t}\right) \mathrm{d}s\,\mathrm{d}t.\end{aligned}$$

The cross product on the right-hand side of this expression is a (not necessarily unital) surface normal determined by the parametrization.

This formula *defines* the integral on the left. We may also interpret this as a special case of integrating 2-forms, where we identify the vector field with a 1-form, and then integrate its Hodge dual over the surface. This is equivalent to integrating $\langle \mathbf{v}, \mathbf{n} \rangle\, \mathrm{d}\Sigma$ over the immersed surface, where $\mathrm{d}\Sigma$ is the induced volume form on the surface, obtained by interior multiplication of the Riemannian metric of the ambient space with the outward normal of the surface.

Surface integrals of Differential 2-forms

Let

$$f = f_z \, \mathrm{d}x \wedge \mathrm{d}y + f_x \, \mathrm{d}y \wedge \mathrm{d}z + f_y \, \mathrm{d}z \wedge \mathrm{d}x$$

be a differential 2-form defined on the surface *S*, and let

$$\mathbf{x}(s,t) = (x(s,t), y(s,t), z(s,t))$$

be an orientation preserving parametrization of *S* with (s,t) in *D*. Changing coordinates from (x,y) to (s,t), the differential forms transform as

$$\mathrm{d}x = \frac{\mathrm{d}x}{\mathrm{d}s}\mathrm{d}s + \frac{\mathrm{d}x}{\mathrm{d}t}\mathrm{d}t$$

$$\mathrm{d}y = \frac{\mathrm{d}y}{\mathrm{d}s}\mathrm{d}s + \frac{\mathrm{d}y}{\mathrm{d}t}\mathrm{d}t$$

So $\mathrm{d}x \wedge \mathrm{d}y$ transforms to $\frac{\partial(x,y)}{\partial(s,t)}\mathrm{d}s \wedge \mathrm{d}t,$, where $\frac{\partial(x,y)}{\partial(s,t)}$ denotes the determinant of the Jacobian of the transition function from (s,t) to (x,y). The transformation of the other forms are similar.

Then, the surface integral of *f* on *S* is given by

$$\iint_D \left[f_z(\mathbf{x}(s,t))\frac{\partial(x,y)}{\partial(s,t)} + f_x(\mathbf{x}(s,t))\frac{\partial(y,z)}{\partial(s,t)} + f_y(\mathbf{x}(s,t))\frac{\partial(z,x)}{\partial(s,t)} \right] \mathrm{d}s\mathrm{d}t$$

where

$$\frac{\partial \mathbf{x}}{\partial s} \times \frac{\partial \mathbf{x}}{\partial t} = \left(\frac{\partial(y,z)}{\partial(s,t)}, \frac{\partial(z,x)}{\partial(s,t)}, \frac{\partial(x,y)}{\partial(s,t)} \right)$$

is the surface element normal to *S*.

Physical Applications of Surface Integrals

Surface integrals are used in multiple areas of physics and engineering. In particular, they are used for calculations of

- mass of a shell;
- center of mass and moments of inertia of a shell;
- gravitational force and pressure force;
- fluid flow and mass flow across a surface;
- electric charge distributed over a surface;
- electric fields (Gauss' Law in electrostatics).

Frullani Integral

We consider integrals of the form

$$I_f(a,b) = \int_0^\infty \frac{f(ax) - f(bx)}{x} dx,$$

where f is a continuous function (real or complex) on (0;1) and a, b > 0. If f(x) tends to a non-zero limit at 0, then the separate integrals of f(ax) / x and f(bx) / x diverge at 0, and a similar comment applies at infinity. The point is that under suitable conditions, the integral of the difference converges.

Theorem. Consider the following conditions:

(C1) $f(x) \to c_0$ as $x \to 0^+$;

(C2) $f(x) \to c_\infty$ as $x \to \infty$;

(C3) there exists K such that $|F(x)| \leq K$ for all $x > 0$, where $F(x) = \int_0^x f(t)\, dt$.

Under conditions (C1) and (C2), we have

$$I_f(a, b) = (c_0 - c_\infty)(\log b - \log a),$$

Under conditions (C1) and (C3), we have

$$I_f(a,b) = c_0(\log b - \log a).$$

Proof. We may assume that b > a: interchanging a and b then gives the case b < a. Let $0 < \delta < X$. The substitution ax = y gives

$$\int_\delta^x \frac{f(ax)}{x} dx = \int_{a\delta}^{aX} \frac{f(y)}{y} dy.$$

So $$\int_\delta^X \frac{f(ax) - f(bx)}{x} dx = \int_{a\delta}^{aX} \frac{f(y)}{y} dy - \int_{b\delta}^{bX} \frac{f(y)}{y} dy$$

$$= \int_{a\delta}^{b\delta} \frac{f(y)}{y} dy - \int_{aX}^{bX} \frac{f(y)}{y} dy$$

$$= I(\delta) - I(X),$$

where we write

$$I(r) = \int_{ar}^{br} \frac{f(y)}{y}\, dy$$

for any r > 0. Let $\varepsilon > 0$ be given. Under condition (C1), there exists $\delta_0 > 0$ such that if $0 < y \leq b\delta_0$, then $|f(y) - c_0| \leq \in$ Then for $\delta \leq \delta_0$, we have $I(\delta) = c_0(\log b - \log a) + r_1(\delta)$,

where

$$r_1(\delta) = \int_{a\delta}^{b\delta} \frac{f(y) - c_0}{y}\, dy,$$

Hence

$$|r_1(\delta)| \leq \int_{a\delta}^{b\delta} \frac{\in}{y}\, dy = \in (\log b - \log a).$$

So $I(\delta) \rightarrow c_0(\log b - \log a)$ as $\delta \rightarrow 0^+$.

In exactly the same way, condition (C2) implies that $I(X) \rightarrow c_\infty(\log b - \log a)$ as $X \rightarrow \infty$.

Under condition (C3), integration by parts gives

$$I(X) = \left[\frac{F(y)}{y}\right]_{aX}^{bX} + \int_{aX}^{bX} \frac{F(y)}{y^2}\, dy,$$

Hence

$$|I(X)| \leq \frac{2K}{aX} + k\int_{aX}^{bX} \frac{1}{y^2}\, dy < \frac{3k}{aX}$$

So $I(X) \rightarrow 0$ as $X \rightarrow 1$. (This also shows that $\int_1^\infty \frac{f(x)}{x}\, dx$ converges; this statement could be taken as the hypothesis instead of (C3)).

In particular, $I_f(1, b)$ equals $(c_0 - c_\infty) \log b$ in case (1), and $c_0 \log b$ in case (2).

We record a number of particular examples which are transparently cases of $I_f(a, b) = (c_0 - c_\infty)(\log b - \log a)$, or $I_f(a,b) = c_0(\log b - \log a)$. without repeatedly writing out the integral expressions:

f(x)	$I_f(a, b)$
e^{-x}	log b - log a
$1/(1 + x^2)$	log b - log a

cos x	log b - log a
e^{-x} cos x	log b - log a
sin x / x	log b - log a
$\tan^{-1}$ x	$\frac{\pi}{2}$ (log a - log b)
tanh x	log a - log b

Among these examples, cos x is the only one satisfying (C3), but not (C2).

We now state a simple extension of the Theorem.

Let G(x) = $\sum_{j=1}^{n} m_j f(a_j x)$, where $a_j > 0$ for $1 \le j \le n$ and $\sum_{j=1}^{n} m_j = 0$

If satises (C1) and (C2), then

$$\int_0^\infty \frac{G(x)}{x}\,dx = (c_\infty - c_0)\sum_{j=1}^{n} m_j \log a_j.$$

If f satisifes (C1) and (C3), the same applies with c_∞ replaced by 0.

Proof. Write $M_j = m_1 + \quad + m_j$. By Abel summation, since $M_n = 0$,

$$G(x) = \sum_{j=1}^{n-1} M_j\left[f(a_j x) - f(a_{j+1}\,x)\right]$$

By (1),

$$\int_0^\infty \frac{G(x)}{x}dx = \left(c_\infty - c_0\right)\sum_{j=1}^{n-1} M_j\,(\log a_j - \log a_{j+1}) = \left(c_\infty - c_0\right)\sum_{j=1}^{n} m_j \log a.$$

Double integral method for (1). For monotonic f, the following is an alternative method for (1) (but not (2)). The method has been in circulation for a long time, at least for the special case $f(x) = e^{-x}$.

Note that for x, y > 0,

$$\frac{1}{y}\frac{d}{dx}f(xy) = \frac{1}{x}\frac{d}{dy}f(xy) = f'(xy).$$

Hence

$$\int_a^b f'(xy)\,dy = \left[\frac{1}{2}f(xy)\right]_{y=a}^{b} = \frac{f(bx) - f(ax)}{x}.$$

Since $f'(x)$ is of constant sign, reversal of the following double integral is justed:

$$-I_f(a,b)=\int_0^\infty\int_a^b f'(xy)dy\,dx=\int_a^b\int_0^\infty f'(xy)dx\,dy.$$

But $\int_0^\infty f'(xy)\,dx=\left[\frac{1}{y}f(xy)\right]_{x=0}^\infty=\frac{1}{y}\left(c_\infty-c_0\right).$

So $-I_f(a,b)=\left(c_\infty-c_0\right)\int_a^b\frac{1}{y}dy=\left(c_\infty-c_0\right)$ (log b - log a).

Some applications of the case $f\left(x\right)=e^{-x}$.

This is undoubtedly the best known Frullani integral. Written out explicitly, the statement is

$$\int_0^\infty\frac{e^{-ax}-e^{-bx}}{x}dx=\log b-\log a$$

The substitution x = -log y transforms (4) to

$$\int_0^1\frac{y^{b-1}-y^{a-1}}{\log y}dy=\log b-\log a.$$

So, for example, $\int_0^1\frac{y-1}{\log y}dy=\log 2$.

Now consider the integral

$$I_1=\int_0^\infty\frac{1}{x^2}(1-e^{-ax})(1-e^{-bx})\,dx.$$

Integration by parts gives $I_1 = A_1 + J_1$, where

$$A_1=\left[-\frac{1}{x}(1-e^{-ax})(1-e^{-bx})\right]_0^\infty$$

$$J_1=\int_0^\infty\frac{1}{x}\left(ae^{-ax}+be^{-bx}-(a+b)e^{-(a+b)x}\right).$$

Since $0<1-e^{-x}<x$ for $x>0$, , we have $\frac{1}{x}(1-e^{-ax})(1-e^{-bx})<abx$, hence $A_1=0$. So by (3),

$$I_1=J_1=(a+b)\log(a+b)-a\log a-b\log b.$$

Next, consider the function

$$E\left(x\right)=\int_x^\infty\frac{e^{-t}}{t}dt.E\ \ x\ \ dx$$

By reversal of the implied double integral, one sees easily that $\int_x^{\infty} E(x)\,dx = 1$. Now let

$$I_2 = \int_0^{\infty} e^{-ax} E(x)\,dx,$$

where a > 0. Reversing the double integral and applying (4), we find.

$$I_2 = \int_0^{\infty} e^{-ax} \int_x^{\infty} \frac{e^{-t}}{t}\,dt\,dx$$

$$= \int_0^{\infty} \frac{e^{-t}}{t}\,dt\,dx \int_0^{t} e^{-ax}\,dx\,dt$$

$$= \int_0^{\infty} \frac{e^{-t}\,(1-e^{-at})}{at}\,dt$$

$$= \frac{1}{a}\log(1+a).$$

Applications of the case $f(x) = \cos x$

We describe two applications of this case. First, since $2 \sin ax \sin bx = \cos(a-b)x - \cos(a+b)x$, we have, for $a > b > 0$,

$$\int_0^{\infty} \frac{\sin ax \sin bx}{x}\,dx = \frac{1}{2}\log\frac{a+b}{a-b}.$$

Second, consider the integral

$$I_3 = \int_0^{\infty} \frac{\sin^5 x}{x^2}\,dx.$$

Since $16 \sin^5 x = 10 \sin x - 5 \sin 3x + \sin 5x$, we have by (3)

$$16\,I_3 = \left[\frac{\sin^5 x}{x}\right]_0^{\infty} + \int_0^{\infty} \frac{1}{x}\left(10 \cos x - 15 \cos 3x + \cos 5x\right)\,dx$$

$$= -\left(10 \log 1 - 15 \log 3 + 5 \log 5\right)$$

$$= 15 \log 3 - 5 \log 5.$$

Antiderivative

Let f be a function defined on an interval I. We say that a function F on I is an antiderivative of f if F is continuous on I and $F'(x) = f(x)$ for all x from the interior Int(I) of I. If such an antiderivative on I exists, then we say that f is (Newton-)integrable on I.

The definition is done in this way because we use the notion of antiderivative in two settings. In order to have a derivative at a certain point x, a function must be defined on a neighborhood of this point. So the most natural situation is when we look for an antiderivative of f on an open interval (a,b). Every point in such an interval has a neighborhood there, so the hoped for antiderivative F can be differentiated everywhere and we have the condition that $F'(x) = f(x)$ on (a,b). Since differentiability implies continuity, such a function F is automatically also continuous on (a,b) = Int(a,b).

However, often we need an antiderivative F on a closed interval $[a,b]$. The definition then specifies how to extend the above natural situation. We still want that $F' = f$ on (a,b) as we cannot have a derivative at endpoints (endpoints have no neighborhood where F would be guaranteed to exist). To describe the required behaviour of F at a and b we use the second condition, that F be continuous on $[a,b]$. By the previous remark we know that due to differentiability, F would be continuous on (a,b) anyway, so the continuity condition actually just adds requirement of one-sided continuity at endpoints of our closed interval.

Example: The function $F(x) = 3x - 1$ is an antiderivative of $f(x) = 3$ on the interval $(0,13]$. Indeed, F is continuous on the real line, so it is also continuous on $(0,13]$. For points from $(0,13)$ = Int$(0,13]$ we also have $[3x-1]' = 3$ as required.

In the same way we check that also $G(x) = 3x + 7$ is an antiderivative of $f(x) = 3$ on $(0,13]$. As a matter of fact, these two functions are antiderivatives to 3 on any interval, typically we would use open or closed ones; the half-open interval in this example was used just to show how it would work.

This example shows several interesting things. First of all, how did we get these antiderivatives? We guessed using our experience with derivatives. Unlike the procedure of differentiation, where we had a reliable algorithm and could differentiate any function that comes our way, the antiderivative is a different story. We will get to it later.

The second interesting thing is that we can have more antiderivatives to the same given function. Again, based on our experience, this should not be surprising; for any constant C, the function $3x + C$ is an antiderivative of $f(x) = 3$ on any given interval.

Are there also any other kinds of antiderivatives of $f(x) = 3$? No. The following theorem says that adding a constant to an already known antiderivative is the only way to get other antiderivatives:

Theorem

Let F be an antiderivative of f on some interval I.

(i) For any constant C, the function $G(x) = F(x) + C$ is an antiderivative of f on I.

(ii) If H is another antiderivative of f on I, then there is a constant C such that $H(x) = F(x) + C$ for all x from I.

The first statement is easy, since G is continuous where F was and

$$G'(x)=\left[F(x)+C\right]' = F'(x)+[C]' = f(x)+0 = f(x).$$

We can also consider this situation from a geometric standpoint. The antiderivative condition $F' = f$ means that we are looking for a function whose graph has tangent lines with prescribed slopes, namely the slope at a point x is equal to $f(x)$. If we find such an F and then add C to it, we are simply moving the graph up or down, therefore the slopes stay the same and the shifted function also satisfies the condition.

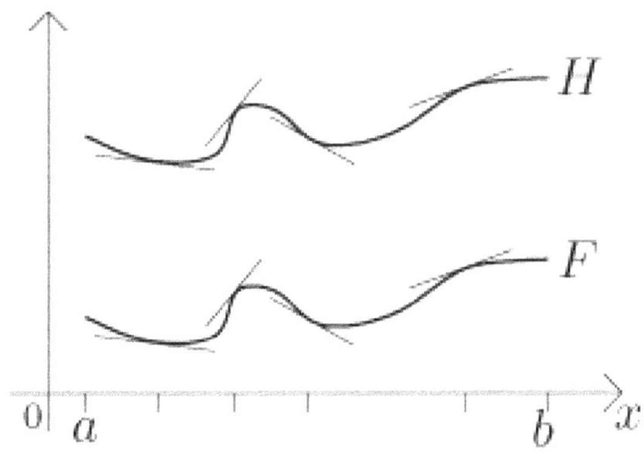

In the converse (ii) we are given a function f prescribing slopes of tangent lines of graphs of two functions, F and H. Imagine that we try to draw these two graphs. We start at their left endpoints and begin drawing both of them simultaneously. Since the two graphs have the same tangent lines at corresponding points, it means that they rise and fall in the same way; that is, we have to move the pencils in a parallel way.

Thus the graphs have to have the same shape, just shifted up or down.

Example

The function $F(x) = x^3/3$ is an antiderivative of $f(x) = x^2$, as the derivative of $x^3/3$ is x^2. As the derivative of a constant is zero, x^2 will have an infinite number of antiderivatives, such as $x^3/3$, $x^3/3 + 1$, $x^3/3 - 2$, etc. Thus, all the antiderivatives of x^2 can be obtained by changing the value of C in $F(x) = x^3/3 + C$; where C is an arbitrary constant known as the constant of integration. Essentially, the graphs of antiderivatives of a given function are vertical translations of each other; each graph's vertical location depending upon the value of C.

In physics, the integration of acceleration yields velocity plus a constant. The constant is the initial velocity term that would be lost upon taking the derivative of velocity because the derivative of a constant term is zero. This same pattern applies to further integrations and derivatives of motion (position, velocity, acceleration, and so on).

Uses and Properties

Antiderivatives are important because they can be used to compute definite integrals, using the fundamental theorem of calculus: if F is an antiderivative of the integrable function f over the interval $[a, b]$, then:

$$\int_a^b f(x)dx = F(b) - F(a).$$

Because of this, each of the infinitely many antiderivatives of a given function f is sometimes called the "general integral" or "indefinite integral" of f and is written using the integral symbol with no bounds:

$$\int f(x)dx.$$

If F is an antiderivative of f, and the function f is defined on some interval, then every other antiderivative G of f differs from F by a constant: there exists a number C such that $G(x) = F(x) + C$ for all x. C is called the constant of integration. If the domain of F is a disjoint union of two or more (open) intervals, then a different constant of integration may be chosen for each of the intervals. For instance

$$F(x) = \begin{cases} -\dfrac{1}{x} + C_1 & x < 0 \\ -\dfrac{1}{x} + C_2 & x > 0 \end{cases}$$

is the most general antiderivative of $f(x) = 1/x^2$ on its natural domain $(-\infty, 0) \cup (0, \infty)$.

Every continuous function f has an antiderivative, and one antiderivative F is given by the definite integral of f with variable upper boundary:

$$F(x) = \int_0^x f(t)dt.$$

Varying the lower boundary produces other antiderivatives (but not necessarily all possible antiderivatives). This is another formulation of the fundamental theorem of calculus.

There are many functions whose antiderivatives, even though they exist, cannot be expressed in terms of elementary functions (like polynomials, exponential functions, logarithms, trigonometric functions, inverse trigonometric functions and their combinations). Examples of these are

$$\int e^{-x^2} dx, \qquad \int \sin x^2\, dx, \qquad \int \frac{\sin x}{x} dx, \qquad \int \frac{1}{\ln x} dx, \qquad \int x^x dx.$$

From left to right, the first four are the error function, the Fresnel function, the trigonometric integral, and the logarithmic integral function.

Techniques of Integration

Finding antiderivatives of elementary functions is often considerably harder than finding their derivatives. For some elementary functions, it is impossible to find an antiderivative in terms of other elementary functions.

There are various methods available:

- the linearity of integration allows us to break complicated integrals into simpler ones
- integration by substitution, often combined with trigonometric identities or the natural logarithm
  - the inverse chain rule method, a special case of integration by substitution
- integration by parts to integrate products of functions
- Inverse function integration, a formula that expresses the antiderivative of the inverse f^{-1} of an invertible and continuous function f in terms of the antiderivative of f and of f^{-1}.
- the method of partial fractions in integration allows us to integrate all rational functions (fractions of two polynomials)
- the Risch algorithm
- when integrating multiple times, certain additional techniques can be used, see for instance double integrals and polar coordinates, the Jacobian and the Stokes' theorem
- if a function has no elementary antiderivative (for instance, exp($-x^2$)), its definite integral can be approximated using numerical integration
- it is often convenient to algebraically manipulate the integrand such that other integration techniques, such as integration by substitution, may be used.
- to calculate the (n times) repeated antiderivative of a function f, Cauchy's formula is useful (cf. Cauchy formula for repeated integration):

$$\int_{x_0}^{x}\int_{x_0}^{x_1}\cdots\int_{x_0}^{x_{n-1}} f(x_n)\,dx_n \ldots dx_2\,dx_1 = \int_{x_0}^{x} f(t)\frac{(x-t)^{n-1}}{(n-1)!}dt.$$

Computer algebra systems can be used to automate some or all of the work involved in the symbolic techniques above, which is particularly useful when the algebraic manipulations involved are very complex or lengthy. Integrals which have already been derived can be looked up in a table of integrals.

Of Non-continuous Functions

Non-continuous functions can have antiderivatives. While there are still open questions in this area, it is known that:

- Some highly pathological functions with large sets of discontinuities may nevertheless have antiderivatives.

- In some cases, the antiderivatives of such pathological functions may be found by Riemann integration, while in other cases these functions are not Riemann integrable.

Assuming that the domains of the functions are open intervals:

- A necessary, but not sufficient, condition for a function *f* to have an antiderivative is that *f* have the intermediate value property. That is, if $[a, b]$ is a subinterval of the domain of *f* and *C* is any real number between $f(a)$ and $f(b)$, then $f(c) = C$ for some *c* between *a* and *b*. This is a consequence of Darboux's theorem.
- The set of discontinuities of *f* must be a meagre set. This set must also be an F-sigma set (since the set of discontinuities of any function must be of this type). Moreover, for any meagre F-sigma set, one can construct some function *f* having an antiderivative, which has the given set as its set of discontinuities.
- If *f* has an antiderivative, is bounded on closed finite subintervals of the domain and has a set of discontinuities of Lebesgue measure 0, then an antiderivative may be found by integration in the sense of Lebesgue. In fact, using more powerful integrals like the Henstock–Kurzweil integral, every function for which an antiderivative exists is integrable, and its general integral coincides with its antiderivative.
- If *f* has an antiderivative *F* on a closed interval $[a,b]$, then for any choice of partition $a = x_0 < x_1 < x_2 < \ldots < x_n = b$, if one chooses sample points $x_i^* \in [x_{i-1}, x_i]$ as specified by the mean value theorem, then the corresponding Riemann sum telescopes to the value $F(b) - F(a)$.

$$\sum_{i=1}^{n} f(x_i^*)(x_i - x_{i-1}) = \sum_{i=1}^{n} [F(x_i) - F(x_{i-1})]$$
$$= F(x_n) - F(x_0) = F(b) - F(a)$$

However if *f* is unbounded, or if *f* is bounded but the set of discontinuities of *f* has positive Lebesgue measure, a different choice of sample points (x_i^*) may give a significantly different value for the Riemann sum, no matter how fine the partition.

Some Examples

1. The function

$$f(x) = 2x\sin\left(\frac{1}{x}\right) - \cos\left(\frac{1}{x}\right)$$

with $f(0) = 0$ is not continuous at $x = 0$ but has the antiderivative

$$F(x) = x^2 \sin\left(\frac{1}{x}\right)$$

with $f(0) = 0$. Since *f* is bounded on closed finite intervals and is only discontinuous at 0, the antiderivative *F* may be obtained by integration: $F(x) = \int_0^x f(t)dt$.

2. The function

$$f(x) = 2x\sin\left(\frac{1}{x^2}\right) - \frac{2}{x}\cos\left(\frac{1}{x^2}\right)$$

with $f(0) = 0$ is not continuous at $x = 0$ but has the antiderivative

$$F(x) = x^2 \sin\left(\frac{1}{x^2}\right)$$

with $F(0) = 0$. Unlike Example, $f(x)$ is unbounded in any interval containing 0, so the Riemann integral is undefined.

3. If $f(x)$ is the function in Example 1 and F is its antiderivative, and $\{x_n\}_{n\geq 1}$ is a dense countable subset of the open interval $(-1,1)$, then the function

$$g(x) = \sum_{n=1}^{\infty} \frac{f(x-x_n)}{2^n}$$

has an antiderivative

$$G(x) = \sum_{n=1}^{\infty} \frac{F(x-x_n)}{2^n}.$$

The set of discontinuities of g is precisely the set $\{x_n\}_{n\geq 1}$. Since g is bounded on closed finite intervals and the set of discontinuities has measure 0, the antiderivative G may be found by integration.

4. Let $\{x_n\}_{n\geq 1}$ be a dense countable subset of the open interval $(-1,1)$. Consider the everywhere continuous strictly increasing function

$$F(x) = \sum_{n=1}^{\infty} \frac{1}{2^n}(x-x_n)^{1/3}.$$

It can be shown that

$$F'(x) = \sum_{n=1}^{\infty} \frac{1}{3\cdot 2^n}(x-x_n)^{-2/3}$$

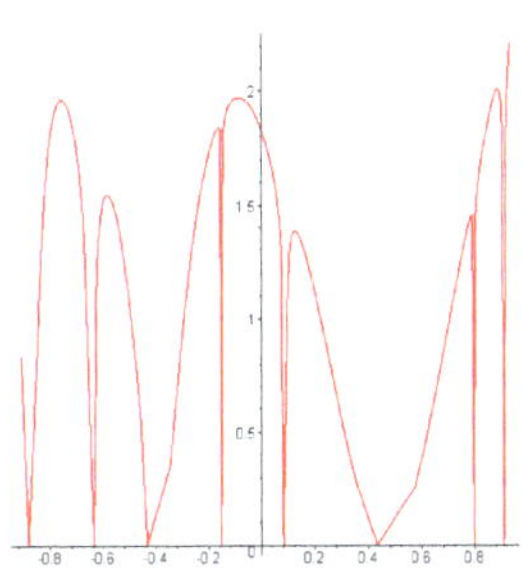

Figure

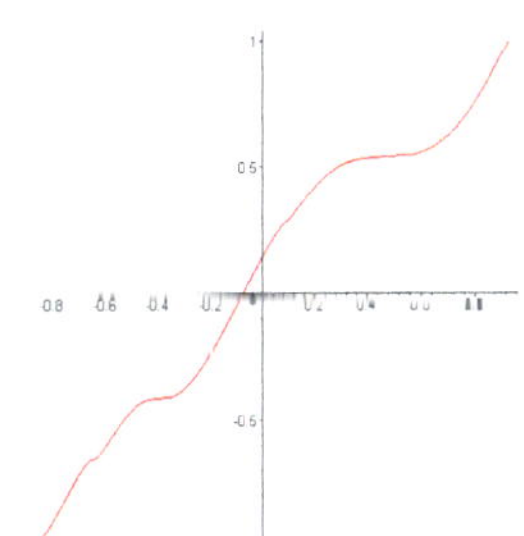

Figure

for all values x where the series converges, and that the graph of $F(x)$ has vertical tangent lines at all other values of x. In particular the graph has vertical tangent lines at all points in the set $\{x_n\}_{n\geq 1}$.

Moreover $F(x) \geq 0$ for all x where the derivative is defined. It follows that the inverse function $G = F^{-1}$ is differentiable everywhere and that

$$g(x) = G'(x) = 0$$

for all x in the set $\{F(x_n)\}_{n\geq 1}$ which is dense in the interval $[F(-1), F(1)]$. Thus g has an antiderivative G. On the other hand, it can not be true that

$$\int_{F(-1)}^{F(1)} g(x)\, dx = GF(1) - GF(-1) = 2,$$

since for any partition of $[F(-1), F(1)]$, one can choose sample points for the Riemann sum from the set $x_n)\}_{n\geq 1}$,, giving a value of 0 for the sum. It follows that g has a set of discontinuities of positive Lebesgue measure. Figure above shows an approximation to the graph of $g(x)$ where $\{x_n = \cos(n)\}_{n\geq 1}$ and the series is truncated to 8 terms. Figure shows the graph of an approximation to the antiderivative $G(x)$, also truncated to 8 terms. On the other hand if the Riemann integral is replaced by the Lebesgue integral, then Fatou's lemma or the dominated convergence theorem shows that g does satisfy the fundamental theorem of calculus in that context.

5. In Examples, the sets of discontinuities of the functions g are dense only in a finite open interval (a, b). However these examples can be easily modified so as to have sets of discontinuities which are dense on the entire real line $(-\infty, \infty)$. Let

$$\lambda(x) = \frac{a+b}{2} + \frac{b-a}{\pi} \tan^{-1} x.$$

Then $g(\lambda(x))\lambda'(x)$ has a dense set of discontinuities on $(-\infty, \infty)$ and has antiderivative $\cdot\lambda$

6. Using a similar method as in Example, one can modify g in Example so as to vanish at all rational numbers. If one uses a naive version of the Riemann integral defined as the limit of left-hand or right-hand Riemann sums over regular partitions, one will obtain that the integral of such a function g over an interval $[a, b]$ is 0 whenever a and b are both rational, instead of $G(b) - G(a)$. Thus the fundamental theorem of calculus will fail spectacularly.

Difference between Antiderivative and Integration

According to the definition of antiderivative and integration, antiderivative is the opposite process of derivative and it was introduced when we discussed the Fundamental theorem of calculus, if the F(x) is the antiderivative of f(x), then the integration of f(x) from a to b equals to the value of F(b) subtract the value of F(a). So basically, the anti-derivative of a function(a) is finding where the function(a) derivative from. In the sense that if function(b) experience derivative can get function(a), so that function(b) is the antiderivative of function(a). Besides the antiderivative is also been called indefinite integrals.

And for the integration, it is a basic and fundamental part of calculus and this includes different types of integrals. We started to learn this from Riemann integration, which is known as definite integral. And this definite integral will end up with a number.

In terms of whether a function have antiderivative or not integral or not, if the function is not continuous they don't have antiderivative. But it may still do integration in terms of definite integral.

Example:

In physics, the time-speed question is a really great example to explain the definitions of antiderivative and integration.

For example:

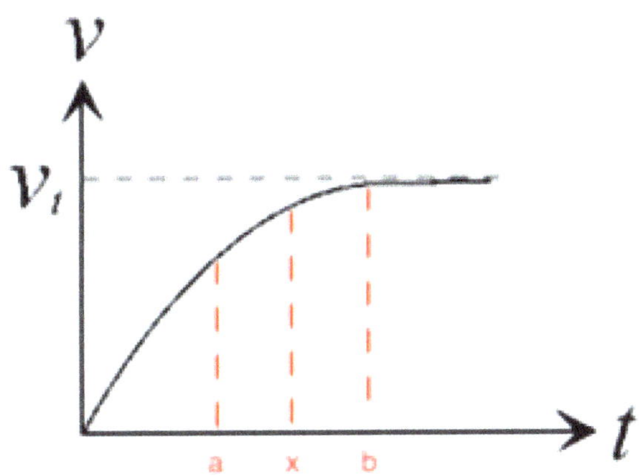

To calculate the distance from time a to b, we need to use the definition of integration. Since a and b are two constants with certain values. From formula d'(t)=v(t),

$$\int_a^b v(t)dt = dt\Big|_a^b = d(b) - d(a).$$

While if we want to express the distance from time a to an uncertain x, the expression of distance is calculated by antiderivative.

Since x is not fixed, the antiderivative of v(t) from time a to x should be d(t). To be more specific, if v(t)=, by the definition of antiderivative , we can get: d(t).

In this way, we can explain the difference between antiderivative and integration.

References

- Stoer, Josef; Bulirsch, Roland (2002), "Chapter 3: Topics in Integration", Introduction to Numerical Analysis (3rd ed.), Springer, ISBN 978-0-387-95452-3
- Hildebrandt, T. H. (1953), "Integration in abstract spaces", Bulletin of the American Mathematical Society, 59 (2): 111–139, doi:10.1090/S0002-9904-1953-09694-X, ISSN 0273-0979
- Rudin, Walter. Principles of Mathematical Analysis. Walter Rudin Student Series in Advanced Mathematics (3rd ed.). McGraw–Hill. ISBN 978-0-07-054235-8
- Lewin, Jonathan (1987). "Some applications of the bounded convergence theorem for an introductory course in analysis". The American Mathematical Monthly. AMS. 94 (10): 988–993. doi:10.2307/2322609
- Kibble, Tom W. B.; Berkshire, Frank H. (2004). Classical Mechanics (5th ed.). Imperial College Press. ISBN 978-1-86094-424-6
- Sinclair, George Edward (1974). "A finitely additive generalization of the Fichtenholz–Lichtenstein theorem". Transactions of the American Mathematical Society. AMS. 193: 359–374. doi:10.2307/1996919

4

Vector Calculus

Vector calculus is a branch of mathematics. It is concerned with the differentiation and integration of vector fields in 3D Euclidean space. The significant concepts of vector operators like the del, gradient, divergence and curl operator have been discussed in detail in this chapter. It also closely examines the key theories of vector calculus such as gradient theorem, divergence theorem, Stokes' theorem and Green's theorem, to provide an extensive understanding of the subject.

Vector calculus (or vector analysis) is a branch of mathematics concerned with differentiation and integration of vector fields. The term "vector calculus" is sometimes used as a synonym for the broader subject of multivariable calculus, which includes vector calculus as well as partial differentiation and multiple integration. Vector calculus plays an important role in differential geometry and in the study of partial differential equations. It is used extensively in physics, especially in the description of electromagnetic and gravitational fields.

Basic Concepts

Fields A field is a continuous function that returns a number (or sets of numbers) for every point in space and time (x, t). There are three basic flavours of fields we will deal with scalar fields A scalar field f(x, y, z, t) returns a single number for every point in space and time. Examples include temperature, salinity, porosity, density. . . . vector fields A vector field F(x, y, z, t) returns a vector for every point in space and is readily visualized as a field of arrows. Examples include velocity, elastic displacements, electric or magnetic fields. tensor fields A second rank tensor field D(x, y, z, t) can be visualized as a field of ellipsoids (3 orthogonal vectors for every point). Examples include stress, strain, strain-rate.

Notation There are many different notations for scalars, vectors and tensors (and they're often mixed and matched); however, a few of the common ones are scalars scalars are usually shown in math italics e.g. f, g, t. . . vectors come in more flavours. when typeset they're usually bold-roman characters e.g. V, or the unit vectors i, j, k. When hand written they usually have a line underneath them. Vectors can also be written in component form as $\mathrm{v} = v_x \mathrm{i} + v_y \mathrm{j} + v_z \mathrm{k}$ or in index notation $\mathrm{v} = v_i \hat{e}_i$ where $\hat{e}_i$ is another representation of the unit vectors.

Tensors (actually second rank tensors) Typeset in sans-serif font D (or often just a bold), handwritten with two underbars, or by component D_{ij} 2nd rank tensors are also conveniently represented by matrices.

Definitions of Basic Operations

Vector Dot Product

$$a \cdot b = a_i b_i = |a||b| \cos \theta$$

is a scalar that records the amount of vector a that lies in the direction of vector b (and vice versa). θ is the smallest angle between the two vectors.

Vector cross product $c = a \times b$ is a vector that is perpendicular to the plane spanned by vectors a and b. The direction that c points in is determined by the right hand rule. Note $a \times b = -b \times a$. The cross product is most easily calculated as the determinant of the matrix

$$c = \begin{vmatrix} i & j & k \\ a_x & a_y & a_z \\ b_x & b_y & b_z \end{vmatrix}$$

Or

$$c = (a_y b_z - a_z b_y)i - (a_x b_z - a_z b_x)j + (a_x b_y - a_y b_x)k$$

or in index notation as $c_i = \epsilon_{ijk} a_j b_k$ where ϵ_{ijk} is the horrid permutation symbol.

Tensor vector dot product is a vector formed by matrix multiplication of a tensor and a vector $c = D \cdot a$. In the case of stress, the force acting on a plane with normal vector n is simply $f = \sigma \cdot n$. Each component of the vector is most easily calculated in index notation with $c_i = D_{ij} a_j$ with summation implied over repeated indices (i.e. $c_1 = D_{11}a_1 + D_{12}a_2 + D_{13}a_3$ and so on for $i = 2,3$.

Vector Algebra

The algebraic (non-differential) operations in vector calculus are referred to as vector algebra, being defined for a vector space and then globally applied to a vector field. The basic algebraic operations consist of:

Operation	Notation	Description
Vector addition	$v_1 + v_2$	Addition of two vector fields, yielding a vector field.
Scalar multiplication	av	Multiplication of a scalar field and a vector field, yielding a vector field.
Dot product	$v_1 \cdot v_2$	Multiplication of two vector fields, yielding a scalar field.
Cross product	$v_1 \times v_2$	Multiplication of two vector fields in $\mathbb{R}^3$, yielding a (pseudo) vector field.

Also commonly used are the two triple products:

Operation	Notation	Description
Scalar triple product	$v_1 \cdot (v_2 \times v_3)$	The dot product of a vector and a cross product of two vectors.
Vector triple product	$v_1 \times (v_2 \times v_3)$	The cross product of a vector and a cross product of two vectors.

Operators and Theorems

Differential Operators

Vector calculus studies various differential operators defined on scalar or vector fields, which are typically expressed in terms of the del operator (∇), also known as "nabla". The three basic vector operators are:

Operation	Notation	Description	Notational analogy	Domain/Range
Gradient	$\text{grad}(f) = \nabla f$	Measures the rate and direction of change in a scalar field.	Scalar multiplication	Maps scalar fields to vector fields.
Divergence	$\text{div}(\mathbf{F}) = \nabla \cdot \mathbf{F}$	Measures the scalar of a source or sink at a given point in a vector field.	Dot product	Maps vector fields to scalar fields.
Curl	$\text{curl}(\mathbf{F}) = \nabla \times \mathbf{F}$	Measures the tendency to rotate about a point in a vector field in $\mathbb{R}^3$.	Cross product	Maps vector fields to (pseudo)vector fields.
f denotes a scalar field and F denotes a vector field				

Also commonly used are the two Laplace operators:

Operation	Notation	Description	Domain/Range
Laplacian	$\Delta f = \nabla^2 f = \nabla \cdot \nabla f$	Measures the difference between the value of the scalar field with its average on infinitesimal balls.	Maps between scalar fields.
Vector Laplacian	$\nabla^2 \mathbf{F} = \nabla(\nabla \cdot \mathbf{F}) - \nabla \times (\nabla \times \mathbf{F})$ $F \quad F \quad f$	Measures the difference between the value of the vector field with its average on infinitesimal balls.	Maps between vector fields.
f denotes a scalar field and F denotes a vector field			

A quantity called the Jacobian matrix is useful for studying functions when both the domain and range of the function are multivariable, such as a change of variables during integration.

Integral Theorems

The three basic vector operators have corresponding theorems which generalize the fundamental theorem of calculus to higher dimensions:

Theorem	Statement	Description
Gradient theorem	$\int_{L\subset\mathbb{R}^n} \nabla\varphi\cdot d\mathbf{r} = \varphi(\mathbf{q})-\varphi(\mathbf{p}) \;\; for \;\; L = L[p\to q]$	The line integral of the gradient of a scalar field over a curve L is equal to the change in the scalar field between the endpoints p and q of the curve.
Divergence theorem	$\underbrace{\int\cdots\int_{V\subset\mathbb{R}^n}}_{n} (\nabla\cdot \mathrm{F})\, dV = \underbrace{\oint\cdots\oint_{\partial V}}_{n-1} \mathrm{F}\cdot d\mathrm{S}$	The integral of the divergence of a vector field over an n-dimensional solid V is equal to the flux of the vector field through the (n−1)-dimensional closed boundary surface of the solid.
Curl (Kelvin–Stokes') theorem	$\iint_{\Sigma\subset\mathbb{R}^3} (\nabla\times \mathrm{F})\cdot d\Sigma = \oint_{\partial\Sigma} \mathrm{F}\cdot dr$	The integral of the curl of a vector field over a surface Σ in $\mathbb{R}^3$ is equal to the circulation of the vector field around the closed curve bounding the surface.
φ denotes a scalar field and F denotes a vector field		

In two dimensions, the divergence and curl theorems reduce to the Green's theorem:

Theorem	Statement	Description
Green's theorem	$\iint_{A\subset\mathbb{R}^2} \left(\frac{\partial M}{\partial x} - \frac{\partial L}{\partial y}\right) dA = \oint_{\partial A} (Ldx + Mdy)$	The integral of the divergence (or curl) of a vector field over some region A in $\mathbb{R}^2$ equals the flux (or circulation) of the vector field over the closed curve bounding the region.
For divergence, $F = (M,-L)$. For curl, $F = (L,M,0)$. L and M are functions of (x, y).		

Applications

Linear Approximations

Linear approximations are used to replace complicated functions with linear functions that are almost the same. Given a differentiable function $f(x,y)$ with real values, one can approximate $f(x,y)$ for (x,y) close to (a,b) by the formula

$$f(x,y) \approx f(a,b) + \frac{\partial f}{\partial x}(a,b)(x-a) + \frac{\partial f}{\partial y}(a,b)(y-b).$$

The right-hand side is the equation of the plane tangent to the graph of $z = f(x,y)$ at (a,b)

Optimization

For a continuously differentiable function of several real variables, a point *P* (that is a set of values for the input variables, which is viewed as a point in R^n) is critical if all of the partial derivatives of the function are zero at *P*, or, equivalently, if its gradient is zero. The critical values are the values of the function at the critical points.

If the function is smooth, or, at least twice continuously differentiable, a critical point may be either a local maximum, a local minimum or a saddle point. The different cases may be distinguished by considering the eigenvalues of the Hessian matrix of second derivatives.

By Fermat's theorem, all local maxima and minima of a differentiable function occur at critical points. Therefore, to find the local maxima and minima, it suffices, theoretically, to compute the zeros of the gradient and the eigenvalues of the Hessian matrix at these zeros.

Physics and Engineering

Vector calculus is particularly useful in studying:

- Center of mass
- Field theory
- Kinematics
- Maxwell's equations

Generalizations

Different 3-Manifolds

Vector calculus is initially defined for Euclidean 3-space, $\mathbb{R}^3$ which has additional structure beyond simply being a 3-dimensional real vector space, namely: a norm (giving a notion of length) defined via an inner product (the dot product), which in turn gives a notion of angle, and an orientation, which gives a notion of left-handed and right-handed. These structures give rise to a volume form, and also the cross product, which is used pervasively in vector calculus.

The gradient and divergence require only the inner product, while the curl and the cross product also requires the handedness of the coordinate system to be taken into account.

Vector calculus can be defined on other 3-dimensional real vector spaces if they have an inner product (or more generally a symmetric nondegenerate form) and an orientation; note that this is less data than an isomorphism to Euclidean space, as it does not require a set of coordinates (a frame of reference), which reflects the fact that vector calculus is invariant under rotations (the special orthogonal group SO(3)).

More generally, vector calculus can be defined on any 3-dimensional oriented Riemannian manifold, or more generally pseudo-Riemannian manifold. This structure simply means that the tangent space at each point has an inner product (more generally, a symmetric nondegenerate form) and an orientation, or more globally that there is a symmetric nondegenerate metric tensor and an orientation, and works because vector calculus is defined in terms of tangent vectors at each point.

Other Dimensions

Most of the analytic results are easily understood, in a more general form, using the machinery of differential geometry, of which vector calculus forms a subset. Grad and div generalize immediately to other dimensions, as do the gradient theorem, divergence theorem, and Laplacian (yielding harmonic analysis), while curl and cross product do not generalize as directly.

From a general point of view, the various fields in (3-dimensional) vector calculus are uniformly seen as being k-vector fields: scalar fields are 0-vector fields, vector fields are 1-vector fields, pseudovector fields are 2-vector fields, and pseudoscalar fields are 3-vector fields. In higher dimensions there are additional types of fields (scalar/vector/pseudovector/pseudoscalar corresponding to $0/1/n-1/n$ dimensions, which is exhaustive in dimension 3), so one cannot only work with (pseudo)scalars and (pseudo)vectors.

In any dimension, assuming a nondegenerate form, grad of a scalar function is a vector field, and div of a vector field is a scalar function, but only in dimension 3 or 7 (and, trivially, in dimension 0 or 1) is the curl of a vector field a vector field, and only in 3 or 7 dimensions can a cross product be defined (generalizations in other dimensionalities either require $n-1$ vectors to yield 1 vector, or are alternative Lie algebras, which are more general antisymmetric bilinear products). The generalization of grad and div, and how curl may be generalized is elaborated at Curl: Generalizations; in brief, the curl of a vector field is a bivector field, which may be interpreted as the special orthogonal Lie algebra of infinitesimal rotations; however, this cannot be identified with a vector field because the dimensions differ – there are 3 dimensions of rotations in 3 dimensions, but 6 dimensions of rotations in 4 dimensions (and more generally $\binom{n}{2} = \frac{1}{2}n(n-1)$ dimensions of rotations in n dimensions).

There are two important alternative generalizations of vector calculus. The first, geometric algebra, uses k-vector fields instead of vector fields (in 3 or fewer dimensions, every k-vector field can be identified with a scalar function or vector field, but this is not true in higher dimensions). This replaces the cross product, which is specific to 3 dimensions, taking in two vector fields and giving as output a vector field, with the exterior product, which exists in all dimensions and takes in two vector fields, giving as output a bivector (2-vector) field. This product yields Clifford algebras as the algebraic structure on vector spaces (with an orientation and nondegenerate form). Geometric algebra is mostly used in generalizations of physics and other applied fields to higher dimensions.

The second generalization uses differential forms (k-covector fields) instead of vector fields or k-vector fields, and is widely used in mathematics, particularly in differential geometry, geometric topology, and harmonic analysis, in particular yielding Hodge theory on oriented pseudo-Riemannian manifolds. From this point of view, grad, curl, and div correspond to the exterior derivative of 0-forms, 1-forms, and 2-forms, respectively, and the key theorems of vector calculus are all special cases of the general form of Stokes' theorem.

From the point of view of both of these generalizations, vector calculus implicitly identifies mathematically distinct objects, which makes the presentation simpler but the underlying mathematical structure and generalizations less clear. From the point of view of geometric algebra, vector calculus implicitly identifies k-vector fields with vector fields or scalar functions: 0-vectors and 3-vectors with scalars, 1-vectors and 2-vectors with vectors. From the point of view of differential forms, vector calculus implicitly identifies k-forms with scalar fields or vector fields: 0-forms and 3-forms with scalar fields, 1-forms and 2-forms with vector fields. Thus for example the curl naturally takes as input a vector field or 1-form, but naturally has as output a 2-vector field or 2-form (hence pseudovector field), which is then interpreted as a vector field, rather than directly taking a vector field to a vector field; this is reflected in the curl of a vector field in higher dimensions not having as output a vector field.

Vector Operator

Three vector calculus operations which find many applications in physics are:

1. The divergence of a vector function

$$\nabla \cdot E = \frac{\partial E_x}{\partial x} + \frac{\partial E_y}{\partial y} + \frac{\partial E_z}{\partial z}$$

2. The curl of a vector function

$$\nabla \times E = \left(\frac{\partial E_z}{\partial y} - \frac{\partial E_y}{\partial z}\right)i + \left(\frac{\partial E_x}{\partial z} - \frac{\partial E_z}{\partial x}\right)j + \left(\frac{\partial E_y}{\partial x} - \frac{\partial E_x}{\partial y}\right)k$$

3. The Gradient of a scalar function

$$\nabla f = \frac{\partial f}{\partial x}i + \frac{\partial f}{\partial y}j + \frac{\partial f}{\partial z}k$$

These examples of vector calculus operations are expressed in Cartesian coordinates, but they can be expressed in terms of any orthogonal coordinate system, aiding in the solution of physical problems which have other than rectangular symmetries.

Del

Vector differential operator `del' is represented by a symbol $\Delta$. Its another name is `nabla'. For three dimensional case, it is defined as

$$\nabla = \text{id/ dx+ jd/ dy+ kd/ dz}$$

where I, j, k are unit vectors in the directions of co-ordinate axis X, Y and Z respectively.

The del operator can be applied to

- a scalar function:

$$\vec{\nabla} z(x,y) \qquad = \vec{e}_x \frac{\partial z}{\partial x} + \vec{e}_y \frac{\partial z}{\partial y}$$

This generates separate values for the two differentials at each point in *x,y* space: $= \vec{e_x} u_x + \vec{e_y} u,$

which can be written as a vector: $= \vec{u}(x,y).$

This *vector function* has an *x* and a y component at each point in (*x,y*) space. The resulting vector function is called the *gradient* of the original scalar function. Thus the electric field is the gradient of the electric potential.

- a vector function:

$$\vec{\nabla}\vec{u}(x,y) \qquad = \vec{e}_x \frac{\partial \vec{u}}{\partial x} + \vec{e}_y \frac{\partial \vec{u}}{\partial y}$$

The differentials apply separately to each component: $= \vec{e}_x \frac{\partial}{\partial x}\left(\vec{e}_x u_x + \vec{e}_y u_y\right) + \vec{e}_y \frac{\partial}{\partial y}\left(\vec{e}_x u_x + \vec{e}_y u_y\right).$

The unit vectors are orthogonal: $= \frac{\partial u_x}{\partial x} + \frac{\partial u_y}{\partial y},$

resulting in a scalar function: $= f(x,y)$

This scalar function $f(x,y)$ is known as the *divergence* of the original vector function.

Hence, the del operator turns scalar functions into vector functions and vector functions into scalar functions. If it is applied twice to a scalar function, the end result is another scalar function:

$$\nabla^2 z(x,y) = \vec{\nabla}\left(\vec{\nabla} z(x,y)\right) = f(x,y)$$

This is the 2nd derivative of $z(x,y)$ *along the steepest slope.*

The Del Operator in Polar Coordinates

In Cartesian coordinates, the del operator takes the same form when applied to scalar and vector functions. This is not the case in polar (spherical or cylindrical) coordinates.

In Cartesian coordinates the three-dimensional del operator is

$$\nabla = \mathrm{i}\frac{\partial}{\partial x} + \mathrm{j}\frac{\partial}{\partial y} + \mathrm{k}\frac{\partial}{z},$$

and the horizontal component is

$$\nabla_H = \mathrm{i}\frac{\partial}{\partial x} + \mathrm{j}\frac{\partial}{\partial y}.$$

Expressions for ∇ in various systems of curvilinear coordinates may be found in any textbook of vector analysis. In meteorology it is often convenient to use a thermodynamic function of state, such as pressure or potential temperature, as the vertical coordinate. If the chosen function is σ, then

$$\nabla = \mathrm{i}\left(\frac{\partial}{\partial x}\right)_\sigma + \mathrm{j}\left(\frac{\partial}{\partial y}\right)_\sigma + \mathrm{k}\left(\frac{\partial \sigma}{\partial z}\right)\frac{\partial}{\partial \sigma},$$

where differentiation with respect to x and y is understood as carried out on surfaces of constant σ (the subscript usually being omitted). The horizontal component is now

$$\nabla_\sigma = \mathrm{i}\frac{\partial}{\partial x} + \mathrm{j}\frac{\partial}{\partial y}.$$

If the quasi-hydrostatic approximation is justified, as in most meteorological contexts, pressure is a useful coordinate, and

$$\nabla = \nabla_p + \nabla_z p\left(\frac{\partial}{\partial p}\right) - \mathrm{k}\, pg\left(\frac{\partial}{\partial p}\right),$$

where g is the acceleration of gravity and ρ is the density. Here

$$\nabla_p = \mathrm{i}\frac{\partial}{\partial x} + \mathrm{j}\frac{\partial}{\partial y},$$

with differentiation carried out in isobaric surfaces, and ∇_z is the horizontal gradient operator in Cartesian-altitude coordinates.

Directional Derivative

The directional derivative of a scalar field $f(x, y, z)$ in the direction $\vec{a}(x, y, z) = a_x\vec{e}_x + a_y\vec{e}_y + a_z\vec{e}_z$ is defined as:

$$\vec{a} \cdot \operatorname{grad} f = a_x\frac{\partial f}{\partial x} + a_y\frac{\partial f}{\partial y} + a_z\frac{\partial f}{\partial z} = \vec{a} \cdot (\nabla f)$$

This gives the rate of change of a field f in the direction of $\vec{a}$. In operator notation, the element in parentheses can be considered a single coherent unit; fluid dynamics uses this convention extensively, terming it the convective derivative—the "moving" derivative of the fluid.

Note that $(\vec{a} \cdot \nabla)$ is an operator that takes scalar to a scalar. It can be extended to operate on a vector, by separately operate on each of its components.

Laplacian

The Laplace operator is a scalar operator that can be applied to either vector or scalar fields; for cartesian coordinate systems it is defined as:

$$\Delta = \frac{\partial^2}{\partial x^2} + \frac{\partial^2}{\partial y^2} + \frac{\partial^2}{\partial z^2} = \nabla \cdot \nabla = \nabla$$

and the definition for more general coordinate systems is given in vector Laplacian.

The Laplacian is ubiquitous throughout modern mathematical physics, appearing for example in Laplace's equation, Poisson's equation, the heat equation, the wave equation, and the Schrödinger equation.

Tensor Derivative

Del can also be applied to a vector field with the result being a tensor. The tensor derivative of a vector field $\vec{v}$ (in three dimensions) is a 9-term second rank tensor – that is, a 3×3 matrix – but can be denoted simply as $\nabla \otimes \vec{v}$, where $\otimes$ represents the dyadic product. This quantity is equivalent to the transpose of the Jacobian matrix of the vector field with respect to space. The divergence of the vector field can then be expressed as the trace of this matrix.

For a small displacement $\delta\vec{r}$, the change in the vector field is given by:

$$\delta\vec{v} = (\nabla \otimes \vec{v}) \cdot \delta\vec{r}$$

Product Rules

For vector calculus:

$$\nabla(fg) = f\nabla g + g\nabla f$$
$$\nabla(\vec{u} \cdot \vec{v}) = \vec{u} \times (\nabla \times \vec{v}) + \vec{v} \times (\nabla \times \vec{u}) + (\vec{u} \cdot \nabla)\vec{v} + (\vec{v} \cdot \nabla)\vec{u}$$
$$\nabla \cdot (f\vec{v}) = f(\nabla \cdot \vec{v}) + \vec{v} \cdot (\nabla f)$$
$$\nabla \cdot (\vec{u} \times \vec{v}) = \vec{v} \cdot (\nabla \times \vec{u}) - \vec{u} \cdot (\nabla \times \vec{v})$$
$$\nabla \times (f\vec{v}) = (\nabla f) \times \vec{v} + f(\nabla \times \vec{v})$$
$$\nabla \times (\vec{u} \times \vec{v}) = \vec{u}(\nabla \cdot \vec{v}) - \vec{v}(\nabla \cdot \vec{u}) + (\vec{v} \cdot \nabla)\vec{u} - (\vec{u} \cdot \nabla)\vec{v}$$

For matrix calculus (for which $\vec{u} \cdot \vec{v}$ can be written $\vec{u}^T \vec{v}$):

$$(\mathbf{A}\nabla)^T \vec{u} = \nabla^T (\mathbf{A}^T \vec{u}) - (\nabla^T \mathbf{A}^T)\vec{u}$$

Another relation of interest is the following, where $\vec{u} \otimes \vec{v}$ is the outer product tensor:

$$\nabla(\vec{u} \otimes \vec{v}) = (\nabla \cdot \vec{u})\vec{v} + (\vec{u} \cdot \nabla)\vec{v}$$

Precautions

Most of the above vector properties (except for those that rely explicitly on del's differential properties—for example, the product rule) rely only on symbol rearrangement, and must necessarily hold if the del symbol is replaced by any other vector. This is part of the value to be gained in notationally representing this operator as a vector.

Though one can often replace del with a vector and obtain a vector identity, making those identities mnemonic, the reverse is *not* necessarily reliable, because del does not commute in general.

A counterexample that relies on del's failure to commute:

$$(\vec{u} \cdot \vec{v})f \equiv (\vec{v} \cdot \vec{u})f$$
$$(\nabla \cdot \vec{v})f = \left(\frac{\partial v_x}{\partial x} + \frac{\partial v_y}{\partial y} + \frac{\partial v_z}{\partial z}\right)f = \frac{\partial v_x}{\partial x}f + \frac{\partial v_y}{\partial y}f + \frac{\partial v_z}{\partial z}f$$
$$(\vec{v} \cdot \nabla)f = \left(v_x\frac{\partial}{\partial x} + v_y\frac{\partial}{\partial y} + v_z\frac{\partial}{\partial z}\right)f = v_x\frac{\partial f}{\partial x} + v_y\frac{\partial f}{\partial y} + v_z\frac{\partial f}{\partial z}$$
$$\Rightarrow (\nabla \cdot \vec{v})f \neq (\vec{v} \cdot \nabla)f$$

A counterexample that relies on del's differential properties:

$$
\begin{aligned}
(\nabla x)\times(\nabla y) &= \left(\vec{e}_x\frac{\partial x}{\partial x}+\vec{e}_y\frac{\partial x}{\partial y}+\vec{e}_z\frac{\partial x}{\partial z}\right)\times\left(\vec{e}_x\frac{\partial y}{\partial x}+\vec{e}_y\frac{\partial y}{\partial y}+\vec{e}_z\frac{\partial y}{\partial z}\right)\\
&= (\vec{e}_x\cdot 1+\vec{e}_y\cdot 0+\vec{e}_z\cdot 0)\times(\vec{e}_x\cdot 0+\vec{e}_y\cdot 1+\vec{e}_z\cdot 0)\\
&= \vec{e}_x\times\vec{e}_y\\
&= \vec{e}_z
\end{aligned}
$$

$$
\begin{aligned}
(\vec{u}x)\times(\vec{u}y) &= xy(\vec{u}\times\vec{u})\\
&= xy\vec{0}\\
&= \vec{0}
\end{aligned}
$$

Central to these distinctions is the fact that del is not simply a vector; it is a vector operator. Whereas a vector is an object with both a magnitude and direction, del has neither a magnitude nor a direction until it operates on a function.

For that reason, identities involving del must be derived with care, using both vector identities and *differentiation* identities such as the product rule.

Gradient

If $f(x, y)$ is a function of two variables x and y, then its gradient at (x, y) is the vector $[\partial f/\partial x, \partial f/\partial y]$. This notion can be defined also in one or three dimensions. In particular, if $f'(x)$ is a function of one variable then its gradient is the one-dimensional vector whose component is $f'(x)$, or in other words the derivative of f. The gradient in higher dimensions is a natural generalization of that, enabling us to calculate the way in which f changes in various directions just as $f'(x)$ measures the change of f in the only direction it could.

Note that the coordinates of the position of a point are written as (x, y), while coordinates of a vector, which measures relative position, are written $[x, y]$.

The gradient also has a natural intrinsic characterization, in that its components measure the rate of change of f in various directions, but in fact we can use it to measure rate of change in any direction. If $\Delta P = [\Delta x, \Delta y]$ is a small change from $P = (x, y)$ then we have approximately

$$
f(P+\Delta P) = f(p) + grad(f)\bullet\Delta P = f(x,y) + (\partial f + \partial x)\Delta x + (\partial f/\partial y)\Delta y.
$$

In other words, the gradient bundles in one package (a vector field) the general idea of rate of change of f.

The Gradient of a Scalar Field

If U(r) = U(x, y, z) is a scalar field, ie a scalar function of position r = [x, y, z] in 3 dimensions, then its gradient at any point is defined in Cartesian co-ordinates by

$$
\text{grad}\,U = \frac{\partial U}{\partial \text{x}}\hat{\text{i}} + \frac{\partial U}{\partial y}\hat{\text{j}} + \frac{\partial U}{\partial z}\hat{\text{k}}
$$

It is usual to define the vector operator which is called "del" or "nabla"

$$\nabla = \hat{i}\frac{\partial}{\partial x} + \hat{j}\frac{\partial}{\partial y} + \hat{k}\frac{\partial}{\partial z}$$

Then

$$\text{grad}\,U \equiv \nabla U$$

Worked examples of gradient evaluation

1. $U = x^2$

$$\Rightarrow \nabla U = \left(\frac{\partial}{\partial x}\hat{i} + \frac{\partial}{\partial y}\hat{j} + \frac{\partial}{\partial z}\hat{k}\right)x^2 = 2x\,\hat{i}.$$

2. $U = r^2$

$$r^2 = x^2 + y^2 + z^2$$

$$\Rightarrow \nabla U = \left(\frac{\partial}{\partial x}\hat{i} + \frac{\partial}{\partial y}\hat{j} + \frac{\partial}{\partial z}\hat{k}\right)\left(x^2 + y^2 + z^2\right)$$

$$= 2x\,\hat{i} + 2y\,\hat{j} + 2z\,\hat{k} = 2\mathbf{r}.$$

3. $U = c \cdot r$, where c is constant.

$$\Rightarrow \nabla U = \left(\hat{i}\frac{\partial}{\partial x} + \hat{j}\frac{\partial}{\partial y} + \hat{k}\frac{\partial}{\partial z}\right)\left(c_1 x + c_2 y + c_3 z\right) = c_1\,\hat{i} + c_2\,\hat{j} + c_3\,\hat{k} = c$$

4. $U = U(r)$, where $r = \sqrt{\left(x^2 + y^2 + z^2\right)}$. NB NOT U(r).

U is a function of r alone so dU/dr exists. As U = U(x, y, z) also,

$$\frac{\partial U}{\partial x} = \frac{dU}{dr}\frac{\partial r}{\partial x} \qquad \frac{\partial U}{\partial y} = \frac{dU}{dr}\frac{\partial r}{\partial y} \qquad \frac{\partial U}{\partial z} = \frac{dU}{dr}\frac{\partial r}{\partial z}.$$

$$\Rightarrow \nabla U = \frac{\partial U}{\partial x}\hat{i} + \frac{\partial U}{\partial y}\hat{j} + \frac{\partial U}{\partial z}\hat{k} \quad = \frac{dU}{dr}\left(\frac{\partial r}{\partial x}\hat{i} + \frac{\partial r}{\partial y}\hat{j} + \frac{\partial r}{\partial z}\hat{k}\right)$$

But $r = \sqrt{\left(x^2 + y^2 + z^2\right)}$, so $\partial r / \partial x = x / r$ and similarly for y, z.

$$\Rightarrow \nabla U = \frac{dU}{dr}\left(\frac{x\,\hat{i} + y\,\hat{j} + z\,\hat{k}}{r}\right) = \frac{dU}{dr}\left(\frac{\mathbf{r}}{r}\right).$$

Cylindrical and Spherical Coordinates

In cylindrical coordinates with a Euclidean metric, the gradient is given by:

$$\nabla f(\rho,\varphi,z)=\frac{\partial f}{\partial \rho}\mathbf{e}_\rho+\frac{1}{\rho}\frac{\partial f}{\partial \varphi}\mathbf{e}_\varphi+\frac{\partial f}{\partial z}\mathbf{e}_z$$

where ρ is the axial distance, φ is the azimuthal or azimuth angle, z is the axial coordinate, and e_ρ, e_φ and e_z are unit vectors pointing along the coordinate directions.

In spherical coordinates, the gradient is given by:

$$\nabla f(r,\theta,\varphi)=\frac{\partial f}{\partial r}\mathbf{e}_r+\frac{1}{r}\frac{\partial f}{\partial \theta}\mathbf{e}_\theta+\frac{1}{r\sin\theta}\frac{\partial f}{\partial \varphi}\mathbf{e}_\varphi$$

where r is the radial distance, φ is the azimuthal angle and θ is the polar angle, and $\mathbf{e}_r$, $\mathbf{e}_\theta$ and $\mathbf{e}_\varphi$ are again local unit vectors pointing in the coordinate directions (i.e. the normalized covariant basis).

General Coordinates

Using Einstein notation we can consider the gradient in general coordinates, which we write as $x^1, \ldots, x^i, \ldots, x^n$, where n is the number of dimensions of the domain. Here, the upper index refers to the number of the coordinate or component, so x^2 refers to the second component, and not the quantity x squared. The index variable i is used to refer to an arbitrary element, such as x^i. The gradient can then be written as:

$$\nabla f=\frac{\partial f}{\partial x^i}\mathbf{e}^i=\frac{\partial f}{\partial x^i}g^{ij}\mathbf{e}_j,$$

where $\mathbf{e}^i$ and $\mathbf{e}_j\left(=\frac{\partial \mathbf{x}}{\partial x^j}\right)$ refer to the unnormalized local contravariant and covariant bases respectively, g^{ij} is the inverse metric tensor, and the Einstein summation convention implies summation over i and j. In terms of the normalized bases, which we refer to as $\hat{\mathbf{e}}^i$ and $\hat{\mathbf{e}}_j$, using the identities $\|\mathbf{e}^i\|=\sqrt{g^{ii}}$ and $\|\mathbf{e}_j\|=\sqrt{g_{jj}}$, we can write:

$$\nabla f=\frac{\partial f}{\partial x^i}\sqrt{g^{ii}}\hat{\mathbf{e}}^i=\frac{\partial f}{\partial x^i}g^{ij}\sqrt{g_{jj}}\hat{\mathbf{e}}_j,$$

which will evaluate to the expressions given above for cylindrical and spherical coordinates.

Gradient and the Derivative or Differential

Linear Approximation to a Function

The gradient of a function f from the Euclidean space $\mathbb{R}^n$ to $\mathbb{R}$ at any particular point x_0 in $\mathbb{R}^n$ characterizes the best linear approximation to f at x_0. The approximation is as follows:

$$f(x)\approx f(x_0)+(\nabla f)_{x_0}\cdot(x-x_0)$$

for x close to x_0, where $(\nabla f)_{x_0}$ is the gradient of f computed at x_0, and the dot denotes the dot product on $\mathbb{R}^n$. This equation is equivalent to the first two terms in the multivariable Taylor series expansion of f at x_0.

Differential or (Exterior) Derivative

The best linear approximation to a differentiable function

$$f : \mathbb{R}^n \to \mathbb{R}$$

at a point x in $\mathbb{R}^n$ is a linear map from $\mathbb{R}^n$ to $\mathbb{R}$ which is often denoted by df_x or $Df(x)$ and called the differential or (total) derivative of f at x. The gradient is therefore related to the differential by the formula

$$(\nabla f)_x \cdot v = df_x(v)$$

for any $v \in \mathbb{R}^n$. The function df, which maps x to df_x, is called the differential or exterior derivative of f and is an example of a differential 1-form.

If $\mathbb{R}^n$ is viewed as the space of (dimension n) column vectors (of real numbers), then one can regard df as the row vector with components

$$\left(\frac{\partial f}{\partial x_1}, \ldots, \frac{\partial f}{\partial x_n}\right)$$

so that $df_x(v)$ is given by matrix multiplication. Assuming the standard Euclidean metric on $\mathbb{R}^n$, the gradient is then the corresponding column vector, i.e.,

$$(\nabla f)_i = df_i^{\mathsf{T}} .$$

Gradient as a Derivative

Let U be an open set in $\mathbb{R}^n$. If the function $f : U \to \mathbb{R}$ is differentiable, then the differential of f is the (Fréchet) derivative of f. Thus ∇f is a function from U to the space $\mathbb{R}$ such that

$$\lim_{h \to 0} \frac{\| f(x+h) - f(x) - \nabla f(x) \cdot h \|}{\| h \|} = 0$$

where $\cdot$ is the dot product.

As a consequence, the usual properties of the derivative hold for the gradient:

Linearity

The gradient is linear in the sense that if f and g are two real-valued functions differentiable at the point $a \in \mathbb{R}^n$, and α and β are two constants, then $\alpha f + \beta g$ is differentiable at a, and moreover

$$\nabla(\alpha f + \beta g)(a) = \alpha \nabla f(a) + \beta \nabla g(a).$$

Product Rule

If f and g are real-valued functions differentiable at a point $a \in \mathbb{R}^n$, then the product rule asserts that the product fg is differentiable at a, and

$$\nabla(fg)(a) = f(a)\nabla g(a) + g(a)\nabla f(a).$$

Chain Rule

Suppose that $f : A \to \mathbb{R}$ is a real-valued function defined on a subset A of $\mathbb{R}^n$, and that f is differentiable at a point a. There are two forms of the chain rule applying to the gradient. First, suppose that the function g is a parametric curve; that is, a function $g : I \to \mathbb{R}^n$ maps a subset $I \subset \mathbb{R}$ into $\mathbb{R}^n$. If g is differentiable at a point $c \in I$ such that $g(c) = a$, then

$$(f \circ g)'(c) = \nabla f(a) \cdot g'(c),$$

where $\circ$ is the composition operator: $(f \circ g)(x) = f(g(x))$.

More generally, if instead $I \subset \mathbb{R}^k$, then the following holds:

$$\nabla(f \circ g)(c) = (Dg(c))^{\mathrm{T}} (\nabla f(a))$$

where $(Dg)^{\mathrm{T}}$ denotes the transpose Jacobian matrix.

For the second form of the chain rule, suppose that $h : I \to \mathbb{R}$ is a real valued function on a subset I of $\mathbb{R}$, and that h is differentiable at the point $f(a) \in I$. Then

$$\nabla(h \circ f)(a) = h'\big(f(a)\big)\nabla f(a).$$

Further Properties and Applications

Level Sets

A level surface, or isosurface, is the set of all points where some function has a given value.

If f is differentiable, then the dot product $(\nabla f)_x \cdot v$ of the gradient at a point x with a vector v gives the directional derivative of f at x in the direction v. It follows that in this case the gradient of f is orthogonal to the level sets of f. For example, a level surface in three-dimensional space is defined by an equation of the form $F(x, y, z) = c$. The gradient of F is then normal to the surface.

More generally, any embedded hypersurface in a Riemannian manifold can be cut out by an equation of the form $F(P) = 0$ such that dF is nowhere zero. The gradient of F is then normal to the hypersurface.

Similarly, an affine algebraic hypersurface may be defined by an equation $F(x_1, \ldots x_n) = 0$, where F is a polynomial. The gradient of F is zero at a singular point of the hypersurface (this is the definition of a singular point). At a non-singular point, it is a nonzero normal vector.

Conservative Vector Fields and the Gradient Theorem

The gradient of a function is called a gradient field. A (continuous) gradient field is always a conservative vector field: its line integral along any path depends only on the endpoints of the path, and can be evaluated by the gradient theorem (the fundamental theorem of calculus for line integrals). Conversely, a (continuous) conservative vector field is always the gradient of a function.

Generalizations

Gradient of a vector

Since the total derivative of a vector field is a linear mapping from vectors to vectors, it is a tensor quantity.

In rectangular coordinates, the gradient of a vector field $\mathbf{f} = (f^1, f^2, f^3)$ is defined by:

$$\nabla \mathbf{f} = g^{jk} \frac{\partial f^i}{\partial x^j} \mathbf{e}_i \otimes \mathbf{e}_k$$

where the Einstein summation notation is used and the tensor product of the vectors $\mathbf{e}_i$ and $\mathbf{e}_k$ is a dyadic tensor of type (2,0), or the transpose of the Jacobian matrix

$$\frac{\partial f^i}{\partial x^j} = \frac{\partial (f^1, f^2, f^3)}{\partial (x^1, x^2, x^3)}.$$

In curvilinear coordinates, or more generally on a curved manifold, the gradient involves Christoffel symbols:

$$\nabla \mathbf{f} = g^{jk} \left(\frac{\partial f^i}{\partial x^j} + \Gamma^i{}_{jl} f^l \right) \mathbf{e}_i \otimes \mathbf{e}_k$$

where g^{jk} are the components of the inverse metric tensor and the $\mathbf{e}_i$ are the coordinate basis vectors.

Expressed more invariantly, the gradient of a vector field $\mathbf{f}$ can be defined by the Levi-Civita connection and metric tensor:

$$\nabla^a f^b = g^{ac} \nabla_c f^b,$$

where ∇_c is the connection.

Riemannian Manifolds

For any smooth function f on a Riemannian manifold (M, g), the gradient of f is the vector field ∇f such that for any vector field X,

$$g(\nabla f, X) = \partial_X f, \qquad i.e., \quad g_x\big((\nabla f)_x, X_x\big) = (\partial_X f)(x)$$

where $g_x(\,,\,)$ denotes the inner product of tangent vectors at x defined by the metric g and $\partial_X f$ is the function that takes any point $x \in M$ to the directional derivative of f in the direction X, evaluated at x. In other words, in a coordinate chart φ from an open subset of M to an open subset of $\mathbb{R}^n$, $(\partial_X f)(x)$ is given by:

$$\sum_{j=1}^{n} X^j(\varphi(x)) \frac{\partial}{\partial x_j} (f \circ \varphi^{-1}) \Big|_{\varphi(x)}$$

where X^j denotes the jth component of X in this coordinate chart.

So, the local form of the gradient takes the form:

$$\nabla f = g^{ik} \frac{\partial f}{\partial x^k} \mathbf{e}_i.$$

Generalizing the case $M = \mathbb{R}^n$, the gradient of a function is related to its exterior derivative, since

$$(\partial_X f)(x) = (df)_x (X_x).$$

More precisely, the gradient ∇f is the vector field associated to the differential 1-form df using the musical isomorphism

$$\sharp = \sharp^g : T^* M \to TM$$

(called "sharp") defined by the metric g. The relation between the exterior derivative and the gradient of a function on $\mathbb{R}^n$ is a special case of this in which the metric is the flat metric given by the dot product.

Divergence

The divergence computes a scalar quantity from a vector field by differentiation.

If a (x, y , z) is a vector function of position in 3 dimensions, that is $a = a_1 \hat{\mathrm{i}} + a_2 \hat{\mathrm{j}} + a_3 \hat{\mathrm{k}}$, then its divergence at any point is defined in Cartesian co-ordinates by

$$\operatorname{div} a = \frac{\partial a_1}{a_x} + \frac{\partial a_2}{\partial_y} + \frac{\partial a_3}{\partial_z}$$

We can write this in a simplified notation using a scalar product with the Δ vector differential operator:

$$\operatorname{div} a = \left(\hat{\mathrm{i}} \frac{\partial}{\partial \mathrm{x}} + \hat{\mathrm{j}} \frac{\partial}{\partial y} + \hat{\mathrm{k}} \frac{\partial}{\partial z} \right) \cdot \mathrm{a} = \nabla \cdot \mathrm{a}$$

Notice that the divergence of a vector field is a scalar field.

Examples of divergence evaluation

	a	div a
1)	$\mathrm{x}\hat{\mathrm{i}}$	1
2)	$\mathrm{r}\left(= \mathrm{x}\hat{\mathrm{i}} + y\hat{\mathrm{j}} + z\hat{\mathrm{k}}\right)$	3
3)	r/r^3	0
4)	rc, for c constant	(r . c)/r

The x component of r/r^3 is $\mathrm{x}.\left(x^2 + y^2 + z^2\right)^{-3/2}$, and we need to find $\partial / \partial \mathrm{x}$ of it.

$$\frac{\partial}{\partial \mathrm{x}} \mathrm{x}.\left(x^2 + y^2 + z^2\right)^{-3/2} = 1.\left(x^2 + y^2 + z^2\right)^{-3/2} + \mathrm{x}\frac{-3}{2}\left(x^2 + y^2 + z^2\right)^{-5/2}.2\mathrm{x}$$

$$= r^{-3}\left(1 - 3\mathrm{x}^2 r^{-2}\right).$$

The terms in y and z are similar, so that

$$\operatorname{div}\left(\mathrm{r}/\mathrm{r}^3\right)=r^{-3}\left(3-3\left(x^2+y^2+z^2\right)r^{-2}\right)=r^{-3}(3-3)$$

$$=0$$

The Significance of Divergence

Consider a typical vector field, water flow, and denote it by a(r). This vector has magnitude equal to the mass of water crossing a unit area perpendicular to the direction of a per unit time.

Now take an infinitesimal volume element dV and figure out the balance of the flow of a in and out of dV.

To be specific, consider the volume element dV = *dx dy dz* in Cartesian co-ordinates, and think first about the face of area dxdz perpendicular to the y axis and facing outwards in the negative y direction. (That is, the one with surface area $d\mathrm{S}=-d\mathrm{x}\ dz\hat{\mathrm{j}}$.)

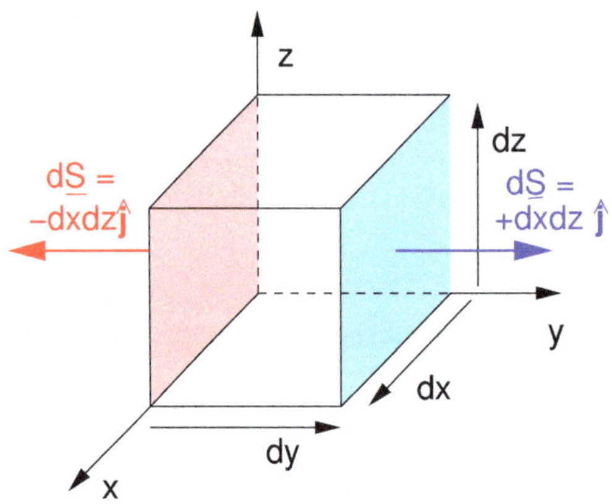

Elemental volume for calculating divergence.

The component of the vector a normal to this face is $\mathrm{a}\cdot\hat{\mathrm{j}}=\mathrm{a}_y$, and is pointing inwards, and so the its contribution to the OUTWARD flux from this surface is

$$\mathrm{a}\cdot d\mathrm{S}=-\mathrm{a}_y(x,y,z)dzd\mathrm{x},$$

(By the way, flux here denotes mass per unit time.)

A similar contribution, but of opposite sign, will arise from the opposite face, but we must remember that we have moved along y by an amount dy, so that this OUTWARD amount is

$$\mathrm{a}_y(\mathrm{x},y+dy,z)dzd\mathrm{x}=\left(\mathrm{a}_y+\frac{\partial\mathrm{a}_y}{\partial y}\right)d\mathrm{x}dz$$

The total outward amount from these two faces is

$$\frac{\partial\mathrm{a}_y}{\partial y}dyd\mathrm{x}dz=\frac{\partial\mathrm{a}_y}{\partial y}dV$$

Summing the other faces gives a total outward flux of

$$\left(\frac{\partial a_x}{\partial x}+\frac{\partial a_y}{\partial y}+\frac{\partial a_z}{\partial z}\right)dV = \nabla\cdot a\, dV$$

The divergence of a vector field represents the flux generation per unit volume at each point of the field. (Divergence because it is an efflux not an influx.)

Physical Interpretation of Divergence

In physical terms, the divergence of a three-dimensional vector field is the extent to which the vector field flow behaves like a source at a given point. It is a local measure of its "outgoingness" – the extent to which there is more of some quantity exiting an infinitesimal region of space than entering it. If the divergence is nonzero at some point then there is compression or expansion at that point.

More rigorously, the divergence of a vector field F at a point p can be defined as the limit of the net flow of F across the smooth boundary of a three-dimensional region V divided by the volume of V as V shrinks to p. Formally,

$$\operatorname{div}\mathbf{F}\Big|_p = \lim_{V\to\{p\}}\iint_{S(V)}\frac{\mathbf{F}\cdot\hat{\mathbf{n}}}{|V|}dS$$

where $|V|$ is the volume of V, $S(V)$ is the boundary of V, and the integral is a surface integral with $\hat{n}$ being the outward unit normal to that surface. The result, div F, is a function of p. From this definition it also becomes obvious that div F can be seen as the *source density* of the flux of F.

In light of the physical interpretation, a vector field with zero divergence everywhere is called *incompressible* or *solenoidal* – in which case any closed surface has no net flow across it.

The intuition that the sum of all sources minus the sum of all sinks should give the net flow outwards of a region is made precise by the divergence theorem.

Cartesian Coordinates

Let x, y, z be a system of Cartesian coordinates in 3-dimensional Euclidean space, and let i, j, k be the corresponding basis of unit vectors. The divergence of a continuously differentiable vector field F = Ui + Vj + Wk is defined as the scalar-valued function:

$$\operatorname{div}\mathbf{F} = \nabla\cdot\mathbf{F} = \left(\frac{\partial}{\partial x},\frac{\partial}{\partial y},\frac{\partial}{\partial z}\right)\cdot(U,V,W) = \frac{\partial U}{\partial x}+\frac{\partial V}{\partial y}+\frac{\partial W}{\partial z}.$$

Although expressed in terms of coordinates, the result is invariant under rotations, as the physical interpretation suggests. This is because the trace of the Jacobian matrix of an N-dimensional vector field F in N-dimensional space is invariant under any invertible linear transformation.

The common notation for the divergence $\nabla \cdot \mathbf{F}$ is a convenient mnemonic, where the dot denotes an operation reminiscent of the dot product: take the components of the ∇ operator, apply them to the corresponding components of F, and sum the results. Because applying an operator is different from multiplying the components, this is considered an abuse of notation.

The divergence of a continuously differentiable second-order tensor field ε is a first-order tensor field:

$$\overrightarrow{\operatorname{div}}(\varepsilon) = \begin{bmatrix} \frac{\partial \varepsilon_{xx}}{\partial x} + \frac{\partial \varepsilon_{yx}}{\partial y} + \frac{\partial \varepsilon_{zx}}{\partial z} \\ \frac{\partial \varepsilon_{xy}}{\partial x} + \frac{\partial \varepsilon_{yy}}{\partial y} + \frac{\partial \varepsilon_{zy}}{\partial z} \\ \frac{\partial \varepsilon_{xz}}{\partial x} + \frac{\partial \varepsilon_{yz}}{\partial y} + \frac{\partial \varepsilon_{zz}}{\partial z} \end{bmatrix}$$

Cylindrical Coordinates

For a vector expressed in local unit cylindrical coordinates as

$$\mathbf{F} = \mathbf{e}_r F_r + \mathbf{e}_\theta F_\theta + \mathbf{e}_z F_z,$$

where e_a is the unit vector in direction a, the divergence is

$$\operatorname{div}\mathbf{F} = \nabla \cdot \mathbf{F} = \frac{1}{r}\frac{\partial}{\partial r}\left(rF_r\right) + \frac{1}{r}\frac{\partial F_\theta}{\partial \theta} + \frac{\partial F_z}{\partial z}.$$

The use of local coordinates is vital for the validity of the expression. If we consider x the position vector and the functions $r(\mathbf{x})$, $\theta(\mathbf{x})$, and $z(\mathbf{x})$, which assign the corresponding global cylindrical coordinate to a vector, in general $r(\mathbf{F}(\mathbf{x})) \neq F_r(\mathbf{x})$, $\theta(\mathbf{F}(\mathbf{x})) \neq F_\theta(\mathbf{x})$, and $z(\mathbf{F}(\mathbf{x})) \neq F_z(\mathbf{x})$. In particular, if we consider the identity function $\mathbf{F}(\mathbf{x}) = \mathbf{x}$, we find that:

$$\theta(\mathbf{F}(\mathbf{x})) = \theta \neq F_\theta(\mathbf{x}) = 0.$$

Spherical Coordinates

In spherical coordinates, with θ the angle with the z axis and φ the rotation around the z axis, and $\mathbf{F}$ again written in local unit coordinates, the divergence is

$$\operatorname{div}\mathbf{F} = \nabla \cdot \mathbf{F} = \frac{1}{r^2}\frac{\partial}{\partial r}\left(r^2 F_r\right) + \frac{1}{r\sin\theta}\frac{\partial}{\partial \theta}(\sin\theta F_\theta) + \frac{1}{r\sin\theta}\frac{\partial F_\varphi}{\partial \varphi}.$$

General Coordinates

Using Einstein notation we can consider the divergence in general coordinates, which we write as $x^1, \ldots, x^i, \ldots, x^n$, where n is the number of dimensions of the domain. Here, the upper index refers to the number of the coordinate or component, so x^2 refers to the second component, and not the quantity x squared. The index variable i is used to refer to an arbitrary element, such as x^i. The divergence can then be written via the Voss- Weyl formula, as:

$$\operatorname{div}(\mathbf{F}) = \frac{1}{\rho}\frac{\partial\left(\rho F^i\right)}{\partial x^i}$$

where ρ is the local coefficient of the volume element and F^i are the components of F with respect to the local unnormalized covariant basis (sometimes written as $\mathbf{e}_i = \frac{\partial \mathbf{x}}{\partial x^i}$). The Einstein notation implies summation over i, since it appears as both an upper and lower index.

The volume coefficient ρ is a function of position which depends on the coordinate system. In Cartesian, cylindrical and polar coordinates, $\rho = 1, r,$ and $r^2 \sin\theta$ respectively, using the same conventions as above. It can also be expressed as $\rho = \sqrt{\det g_{ab}}$, where g_{ab} is the metric tensor. Since the determinant is a scalar quantity which doesn't depend on the indices, we can suppress them and simply write $\rho = \sqrt{\det g}$. Another expression comes from computing the determinant of the Jacobian for transforming from Cartesian coordinates, which for n=3 gives $\rho = \left|\frac{\partial(x,y,z)}{\partial(x^1,x^2,x^3)}\right|$.

Some conventions expect all local basis elements to be normalized to unit length. If we write $\hat{\mathbf{e}}_i$ for the normalized basis, and $\hat{F}^i$ for the components of F with respect to it, we have that $\mathbf{F} = F^i\mathbf{e}_i = F^i \|\mathbf{e}_i\| \frac{\mathbf{e}_i}{\|\mathbf{e}_i\|} = F^i\sqrt{g_{ii}}\,\hat{\mathbf{e}}_i = \hat{F}^i\hat{\mathbf{e}}_i$, using one of the properties of the metric tensor. By dotting both sides of the last equality with the contravariant element $\hat{\mathbf{e}}^i$, we can conclude that $F^i = \frac{\hat{F}^i}{\sqrt{g_{ii}}}$. After substituting, the formula becomes:

$$\operatorname{div}(\mathbf{F}) = \frac{1}{\rho}\frac{\partial\left(\frac{\rho}{\sqrt{g_{ii}}}\hat{F}^i\right)}{\partial x^i} = \frac{1}{\sqrt{\det g}}\frac{\partial\left(\sqrt{\frac{\det g}{g_{ii}}}\hat{F}^i\right)}{\partial x^i}.$$

Decomposition Theorem

It can be shown that any stationary flux v(r) that is at least twice continuously differentiable in $\mathbb{R}^3$ and vanishes sufficiently fast for |r| → ∞ can be decomposed into an *irrotational part* E(r) and a *source-free part* B(r). Moreover, these parts are explicitly determined by the respective *source densities* and *circulation densities*:

For the irrotational part one has

$$\mathbf{E} = -\nabla\Phi(\mathbf{r}),$$

with

$$\Phi(\mathbf{r}) = \int_{\mathbb{R}^3} d^3\mathbf{r}' \frac{\operatorname{div}\mathbf{v}(\mathbf{r}')}{4\pi|\mathbf{r}-\mathbf{r}'|}.$$

The source-free part, B, can be similarly written: one only has to replace the *scalar potential* Φ(r) by a *vector potential* A(r) and the terms $-\nabla\Phi$ by $+\nabla \times A$, and the source density div v by the circulation density $\nabla \times v$.

This "decomposition theorem" is a by-product of the stationary case of electrodynamics. It is a special case of the more general Helmholtz decomposition which works in dimensions greater than three as well.

Properties

The following properties can all be derived from the ordinary differentiation rules of calculus. Most importantly, the divergence is a linear operator, i.e.

$$\operatorname{div}(a\mathbf{F} + b\mathbf{G}) = a \operatorname{div}\mathbf{F} + b \operatorname{div}\mathbf{G}$$

for all vector fields **F** and **G** and all real numbers *a* and *b*.

There is a product rule of the following type: if φ is a scalar-valued function and F is a vector field, then

$$\operatorname{div}(\varphi\mathbf{F}) = \operatorname{grad}\varphi \cdot \mathbf{F} + \varphi \operatorname{div}\mathbf{F},$$

or in more suggestive notation

$$\nabla \cdot (\varphi\mathbf{F}) = (\nabla\varphi) \cdot \mathbf{F} + \varphi(\nabla \cdot \mathbf{F}).$$

Another product rule for the cross product of two vector fields F and G in three dimensions involves the curl and reads as follows:

$$\operatorname{div}(\mathbf{F} \times \mathbf{G}) = \operatorname{curl}\mathbf{F} \cdot \mathbf{G} - \mathbf{F} \cdot \operatorname{curl}\mathbf{G},$$

or

$$\nabla \cdot (\mathbf{F} \times \mathbf{G}) = (\nabla \times \mathbf{F}) \cdot \mathbf{G} - \mathbf{F} \cdot (\nabla \times \mathbf{G}).$$

The Laplacian: Div (GradU) of a Scalar Field

Recall that gradU of any scalar field U is a vector field. Recall also that we can compute the divergence of any vector field. So we can certainly compute div (gradU), even if we don't know what it means yet.

Here is where the ∇ operator starts to be really handy.

$$\nabla.(\nabla U) = \left(\hat{i}\frac{\partial}{\partial x} + \hat{j}\frac{\partial}{\partial y} + \hat{k}\frac{\partial}{\partial z}\right) \cdot \left(\left(\hat{i}\frac{\partial}{\partial x} + \hat{j}\frac{\partial}{\partial y} + \hat{k}\frac{\partial}{\partial z}\right)U\right)$$

$$= \left(\left(\hat{i}\frac{\partial}{\partial x} + \hat{j}\frac{\partial}{\partial y} + \hat{k}\frac{\partial}{\partial z}\right) \cdot \left(\hat{i}\frac{\partial}{\partial x} + \hat{j}\frac{\partial}{\partial y} + \hat{k}\frac{\partial}{\partial z}\right)\right)U$$

$$=\left(\frac{\partial^2}{\partial x^2}+\frac{\partial^2}{\partial y^2}+\frac{\partial^2}{\partial z^2}\right)U$$

$$=\left(\frac{\partial^2 U}{\partial x^2}+\frac{\partial^2 U}{\partial y^2}+\frac{\partial^2 U}{\partial z^2}\right)$$

This last expression occurs frequently in engineering science (you will meet it next in solving Laplace's Equation in partial differential equations). For this reason, the operator ∇^2 is called the "Laplacian"

$$\nabla^2 U\left(\frac{\partial^2}{\partial x^2}+\frac{\partial^2}{\partial y^2}+\frac{\partial^2}{\partial z^2}\right)U$$

Laplace's equation itself is

$$\nabla^2 U = 0$$

Examples of $\nabla^2 U$ evaluation

U	$\nabla^2 U$
1) $r^2\left(=x^2+y^2+z^2\right)$	6
2) xy^2z^3	$2xz^3+6xy^2z$
3) $1/r$	0

Let's prove example

$$1/r=\left(x^2+y^2+z^2\right)^{-1/2}$$

$$\frac{\partial}{\partial x}\frac{\partial}{\partial x}\left(x^2+y^2+z^2\right)^{-1/2}=\frac{\partial}{\partial x}-x.\left(x^2+y^2+z^2\right)^{-3/2}$$

$$=-\left(x^2+y^2+z^2\right)^{-3/2}+3x.x.\left(x^2+y^2+z^2\right)^{-5/2}$$

$$=\left(1/r^3\right)\left(-1+3x^2/r^2\right)$$

Adding up similar terms for y and z

$$\nabla^2\frac{1}{r}=\frac{1}{r^3}\left(-3+3\frac{\left(x^2+y^2+x^2\right)}{r^2}\right)=0$$

Relation with the Exterior Derivative

One can express the divergence as a particular case of the exterior derivative, which takes a 2-form to a 3-form in $\mathbb{R}^3$. Define the current two-form as

$$j=F_1\,dy\wedge dz+F_2\,dz\wedge dx+F_3\,dx\wedge dy.$$

It measures the amount of "stuff" flowing through a surface per unit time in a "stuff fluid" of density $\rho = 1\, dx \wedge dy \wedge dz$ moving with local velocity F. Its exterior derivative dj is then given by

$$dj = \left(\frac{\partial F_1}{\partial x} + \frac{\partial F_2}{\partial y} + \frac{\partial F_3}{\partial z} \right) dx \wedge dy \wedge dz = (\nabla \cdot \mathbf{F})\rho.$$

Thus, the divergence of the vector field F can be expressed as:

$$\nabla \cdot \mathbf{F} = \star d \star \left(\mathbf{F}^{\flat} \right).$$

Here the superscript ♭ is one of the two musical isomorphisms, and ⋆ is the Hodge dual. Working with the current two-form and the exterior derivative is usually easier than working with the vector field and divergence, because unlike the divergence, the exterior derivative commutes with a change of (curvilinear) coordinate system.

Generalizations

The divergence of a vector field can be defined in any number of dimensions. If

$$\mathbf{F} = (F_1, F_2, \ldots F_n),$$

in a Euclidean coordinate system with coordinates $x_1, x_2, \ldots, x_n$, define

$$\operatorname{div} \mathbf{F} = \nabla \cdot \mathbf{F} = \frac{\partial F_1}{\partial x_1} + \frac{\partial F_2}{\partial x_2} + \cdots + \frac{\partial F_n}{\partial x_n}.$$

The appropriate expression is more complicated in curvilinear coordinates.

In the case of one dimension, F reduces to a regular function, and the divergence reduces to the derivative.

For any n, the divergence is a linear operator, and it satisfies the "product rule"

$$\nabla \cdot (\varphi \mathbf{F}) = (\nabla \varphi) \cdot \mathbf{F} + \varphi (\nabla \cdot \mathbf{F})$$

for any scalar-valued function φ.

The divergence of a vector field extends naturally to any differentiable manifold of dimension n that has a volume form (or density) μ, e.g. a Riemannian or Lorentzian manifold. Generalising the construction of a two-form for a vector field on $\mathbb{R}^3$, on such a manifold a vector field X defines an $(n - 1)$-form $j = i_X \mu$ obtained by contracting X with μ. The divergence is then the function defined by

$$dj = (\operatorname{div} X)\mu.$$

Standard formulas for the Lie derivative allow us to reformulate this as

$$\mathcal{L}_X \mu = (\operatorname{div} X)\mu.$$

This means that the divergence measures the rate of expansion of a volume element as we let it flow with the vector field.

On a pseudo-Riemannian manifold, the divergence with respect to the metric volume form can be computed in terms of the Levi-Civita connection ∇:

$$\operatorname{div} X = \nabla \cdot X = X^a{}_{;a},$$

where the second expression is the contraction of the vector field valued 1- form ∇X with itself and the last expression is the traditional coordinate expression from Ricci calculus.

An equivalent expression without using connection is

$$\operatorname{div}(X) = \frac{1}{\sqrt{\det g}} \partial_a \left(\sqrt{\det g}\, X^a \right),$$

where g is the metric and ∂_a denotes the partial derivative with respect to coordinate x^a.

Divergence can also be generalised to tensors. In Einstein notation, the divergence of a contravariant vector F^μ is given by

$$\nabla \cdot \mathbf{F} = \nabla_\mu F^\mu,$$

where ∇_μ denotes the covariant derivative.

Equivalently, some authors define the divergence of a mixed tensor by using the musical isomorphism ♯: if T is a (p, q)-tensor (p for the contravariant vector and q for the covariant one), then we define the *divergence of T* to be the $(p, q - 1)$-tensor

$$(\operatorname{div} T)(Y_1, \ldots, Y_{q-1}) = \operatorname{trace}\left(X \mapsto \sharp(\nabla T)(X, \cdot, Y_1, \ldots, Y_{q-1}) \right);$$

that is, we take the trace over the *first two* covariant indices of the covariant derivative.

Curl

The curl of a vector field A, denoted by curl A or $\nabla \times A$, is a vector whose magnitude is the maximum net circulation of A per unit area as the area tends to zero and whose direction is the normal direction of the area when the area is oriented to make the net circulation maximum.

Cross it with a vector field $\nabla \times \mathrm{a}$

This gives the curl of a vector field

$$\nabla \times \mathrm{a} \equiv \operatorname{curl}(\mathrm{a})$$

We can follow the pseudo-determinant recipe for vector products, so that

$$\nabla \times \mathrm{a} = \begin{vmatrix} \hat{\mathrm{i}} & \hat{\mathrm{j}} & \hat{\mathrm{k}} \\ \dfrac{\partial}{\partial \mathrm{x}} & \dfrac{\partial}{\partial y} & \dfrac{\partial}{\partial z} \\ \partial \mathrm{x} & \partial y & \partial z \end{vmatrix}$$

$$= \left(\frac{\partial \mathrm{a}_z}{\partial y} - \frac{\partial \mathrm{a}_y}{\partial z} \right) \hat{\mathrm{i}} + \left(\frac{\partial \mathrm{a}_\mathrm{x}}{\partial z} - \frac{\partial \mathrm{a}_z}{\partial \mathrm{x}} \right) \hat{\mathrm{j}} + \left(\frac{\partial \mathrm{a}_y}{\partial \mathrm{x}} - \frac{\partial \mathrm{a}_\mathrm{x}}{\partial y} \right) \hat{\mathrm{k}}$$

Examples of Curl Evaluation

	a	$\nabla \times a$
1)	$-y\hat{i} + x\hat{j}$	$2\hat{k}$
2)	$x^2 y^2 \hat{k}$	$2x^2 y\hat{i}' - 2x y^2 \hat{j}$

The Signficance of Curl

Perhaps the first example gives a clue. The field $a = -y\hat{i} + x\hat{j}$ is sketched in Figure. (It is the field you would calculate as the velocity field of an object rotating with $\dot{u} = [0,0,1]$.) This field has a curl of $2\hat{k}$, which is in the r-h screw sense out of the page. We can also see that a field like this must give a finite value to the line integral around the complete loop $\oint_C a \cdot dr$.

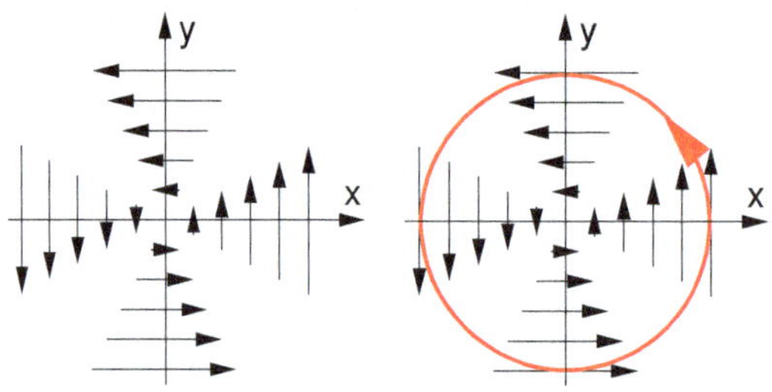

A rough sketch of the vector field $-y\hat{i} + x\hat{j}$.

In fact curl is closely related to the line integral around a loop.

The circulation of a vector a round any closed curve C is defined to be $\oint_C a \cdot dr$ and the curl of the vector field a represents the vorticity, or circulation per unit area, of the field.

Our proof uses the small rectangular element dx by dy shown in Figure.

Consider the circulation round the perimeter of a rectangular element.

The fields in the x direction at the bottom and top are:

$$a_x(x,y,z) \quad \text{and} \quad a_x(x,y+dy,z) = a_x(x,y,z) + \frac{\partial a_x}{\partial y} dy,$$

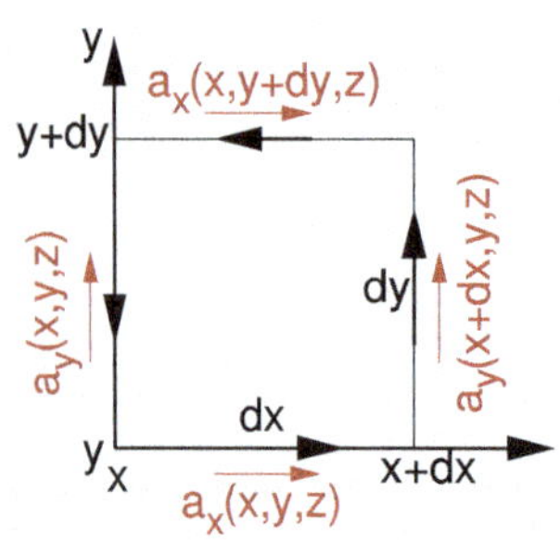

A small element used to calculate curl.

and the fields in the y direction at the left and right are

$$a_y(x,y,z) \quad \text{and} \quad a_y(x+dx,y,z) = a_y(x,y,z) + \frac{\partial a_y}{\partial x}dx$$

Starting at the bottom and working round in the anticlockwise sense, the four contributions to the circulation dC are therefore as follows, where the minus signs take account of the path being opposed to the field:

$$\begin{aligned} dC &= +[a_x\, dx] + [a_y(x+dx,y,z)dy] - [a_x(x,y+dy,z)dx] - [ay\, dy] \\ &= +[a_x\, dx] + \left[\left(a_y + \frac{\partial a_y}{\partial x}dx\right)dy\right] - \left[\left(a_x(y) + \frac{\partial a_x}{\partial y}\right)dx\right] - [a_y\, dy] \\ &= \left(\frac{\partial a_y}{\partial x} - \frac{\partial a_x}{\partial y}\right)dx\, dy \\ &= (\nabla \times a)\cdot dS \end{aligned}$$

Where $dS = dxdy\hat{k}$.

Usage

In practice, the above definition is rarely used because in virtually all cases, the curl operator can be applied using some set of curvilinear coordinates, for which simpler representations have been derived.

The notation $\nabla \times F$ has its origins in the similarities to the 3-dimensional cross product, and it is useful as a mnemonic in Cartesian coordinates if ∇ is taken as a vector differential operator del. Such notation involving operators is common in physics and algebra. However, in certain coordinate systems, such as polar-toroidal coordinates (common in plasma physics), using the notation $\nabla \times F$ will yield an incorrect result.

Expanded in Cartesian coordinates, $\nabla \times F$ is, for F composed of $[F_x, F_y, F_z]$:

$$\begin{vmatrix} \mathrm{i} & \mathbf{j} & \mathbf{k} \\ \frac{\partial}{\partial x} & \frac{\partial}{\partial y} & \frac{\partial}{\partial z} \\ F_x & F_y & F_z \end{vmatrix}$$

where i, j, and k are the unit vectors for the *x*-, *y*-, and *z*-axes, respectively. This expands as follows:

$$\left(\frac{\partial F_z}{\partial y} - \frac{\partial F_y}{\partial z}\right)\mathbf{i} + \left(\frac{\partial F_x}{\partial z} - \frac{\partial F_z}{\partial x}\right)\mathbf{j} + \left(\frac{\partial F_y}{\partial x} - \frac{\partial F_x}{\partial y}\right)\mathbf{k}$$

Although expressed in terms of coordinates, the result is invariant under proper rotations of the coordinate axes but the result inverts under reflection.

In a general coordinate system, the curl is given by

$$(\nabla \times \mathbf{F})^k = \varepsilon^{k\ell m}\nabla_\ell F_m$$

where ε denotes the Levi-Civita tensor and ∇ the covariant derivative, the metric tensor is used to lower the index on F, and the Einstein summation convention implies that repeated indices are summed over. Equivalently,

$$(\nabla \times \mathbf{F}) = \mathbf{e}_k \varepsilon^{k\ell m} \nabla_l F_m$$

where $\mathbf{e}_k$ are the coordinate vector fields. Equivalently, using the exterior derivative, the curl can be expressed as:

$$\nabla \times \mathbf{F} = \left(\star(\mathbf{d}F^\flat)\right)^\sharp$$

Here ♭ and ♯ are the musical isomorphisms, and ★ is the Hodge dual. This formula shows how to calculate the curl of F in any coordinate system, and how to extend the curl to any oriented three-dimensional Riemannian manifold. Since this depends on a choice of orientation, curl is a chiral operation. In other words, if the orientation is reversed, then the direction of the curl is also reversed.

Example

Take the vector field:

$$\mathbf{F}(x, y, z) = y\hat{\mathbf{x}} - x\hat{\mathbf{y}}.$$

Its corresponding plot:

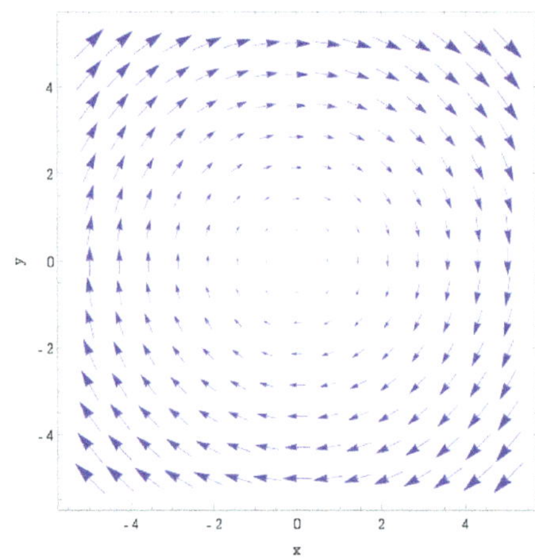

Upon visual inspection, the field can be described as "rotating". If a stationary object were to be placed in the field with the vectors representing a linear force, the object would rotate clockwise.

Calculating the curl:

$$\nabla \times \mathbf{F} = 0\hat{\mathbf{x}} + 0\hat{\mathbf{y}} + \left(\frac{\partial}{\partial x}(-x) - \frac{\partial}{\partial y} y\right)\hat{\mathbf{z}} = -2\hat{\mathbf{z}}$$

The resulting vector field describing the curl would be uniformly going in the negative z direction. It should be noted that the results of this equation align with what could have been predicted using the right-hand rule using a right-handed coordinate system. Being a uniform vector field, the object described before would have the same rotational intensity regardless of where it was placed.

The plot describing the curl of F:

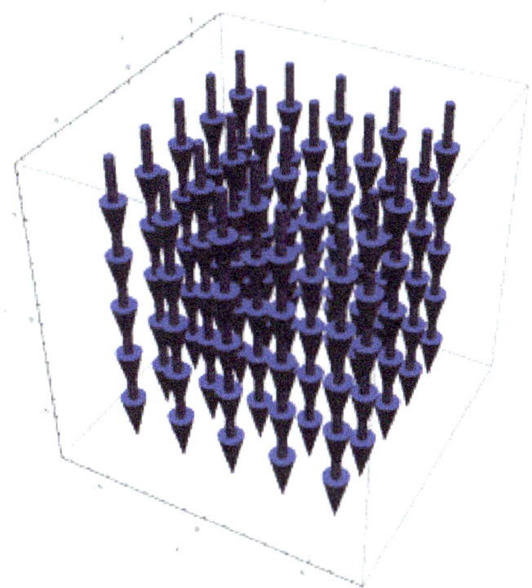

Take the vector field:

$$\mathbf{F}(x, y, z) = -x^2\hat{\mathbf{y}}.$$

Its corresponding plot:

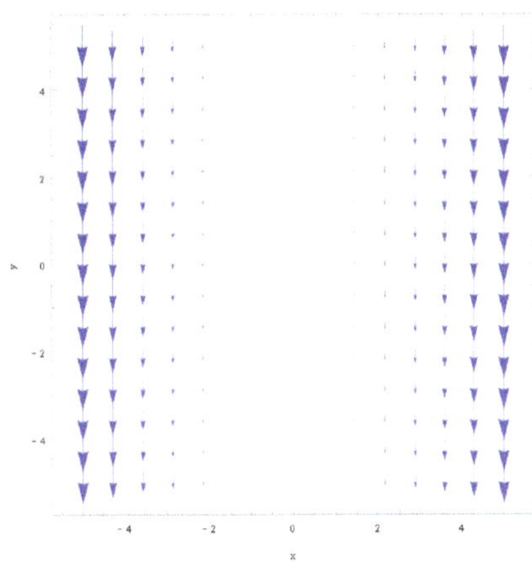

Upon initial inspection, curl existing in this graph would not be obvious. However, taking the object in the previous example, and placing it anywhere on the line $x = 3$, the force exerted on the right side would be slightly greater than the force exerted on the left, causing it to rotate clockwise. Using the right-hand rule, it can be predicted that the resulting curl would be straight in the negative z direction. Inversely, if placed on $x = -3$, the object would rotate counterclockwise and the right-hand rule would result in a positive z direction.

Calculating the curl:

$$\nabla \times \mathbf{F} = 0\hat{\mathbf{x}} + 0\hat{\mathbf{y}} + \frac{\partial}{\partial x}(-x^2)\hat{\mathbf{z}} = -2x\hat{\mathbf{z}}.$$

As predicted, the curl points in the negative z direction when x is positive and vice versa. In this field, the intensity of rotation would be greater as the object moves away from the plane $x = 0$.

The plot describing the curl of **F**:

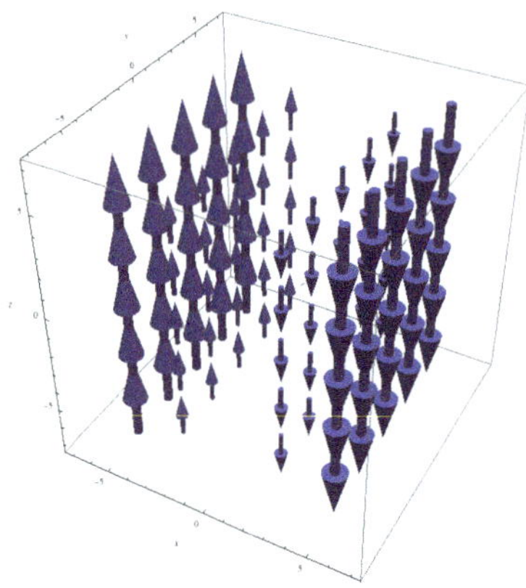

Identities

In general curvilinear coordinates (not only in Cartesian coordinates), the curl of a cross product of vector fields **v** and **F** can be shown to be

$$\nabla \times (\mathbf{v} \times \mathbf{F}) = \big((\nabla \cdot \mathbf{F}) + \mathbf{F} \cdot \nabla\big)\mathbf{v} - \big((\nabla \cdot \mathbf{v}) + \mathbf{v} \cdot \nabla\big)\mathbf{F}$$

Interchanging the vector field **v** and ∇ operator, we arrive at the cross product of a vector field with curl of a vector field:

$$\mathbf{v} \times (\nabla \times \mathbf{F}) = \nabla_{\mathbf{F}}(\mathbf{v} \cdot \mathbf{F}) - (\mathbf{v} \cdot \nabla)\mathbf{F},$$

where $\nabla_{\mathbf{F}}$ is the Feynman subscript notation, which considers only the variation due to the vector field **F** (i.e., in this case, **v** is treated as being constant in space).

Another example is the curl of a curl of a vector field. It can be shown that in general coordinates

$$\nabla \times (\nabla \times \mathbf{F}) = \nabla(\nabla \cdot \mathbf{F}) - \nabla^2 \mathbf{F},$$

and this identity defines the vector Laplacian of **F**, symbolized as $\nabla^2\mathbf{F}$.

The curl of the gradient of *any* scalar field φ is always the zero vector field

$$\nabla \times (\nabla \varphi) = \mathbf{0}$$

which follows from the antisymmetry in the definition of the curl, and the symmetry of second derivatives.

If φ is a scalar valued function and **F** is a vector field, then

$$\nabla \times (\varphi \mathbf{F}) = \nabla \varphi \times \mathbf{F} + \varphi \nabla \times \mathbf{F}$$

Descriptive Examples

- In a vector field describing the linear velocities of each part of a rotating disk, the curl has the same value at all points.

- Of the four Maxwell's equations, two—Faraday's law and Ampère's law—can be compactly expressed using curl. Faraday's law states that the curl of an electric field is equal to the opposite of the time rate of change of the magnetic field, while Ampère's law relates the curl of the magnetic field to the current and rate of change of the electric field.

Generalizations

The vector calculus operations of grad, curl, and div are most easily generalized and understood in the context of differential forms, which involves a number of steps. In a nutshell, they correspond to the derivatives of 0-forms, 1-forms, and 2-forms, respectively. The geometric interpretation of curl as rotation corresponds to identifying bivectors (2-vectors) in 3 dimensions with the special orthogonal Lie algebra $\mathfrak{so}(3)$ of infinitesimal rotations (in coordinates, skew-symmetric 3×3 matrices), while representing rotations by vectors corresponds to identifying 1-vectors (equivalently, 2-vectors) and $\mathfrak{so}(3)$, these all being 3-dimensional spaces.

Differential Forms

In 3 dimensions, a differential 0-form is simply a function $f(x, y, z)$; a differential 1-form is the following expression:

$$a_1\,dx + a_2\,dy + a_3\,dz;$$

a differential 2-form is the formal sum:

$$a_{12}\,dx \wedge dy + a_{13}\,dx \wedge dz + a_{23}\,dy \wedge dz;$$

and a differential 3-form is defined by a single term:

$$a_{123}\,dx \wedge dy \wedge dz.$$

(Here the a-coefficients are real functions; the "wedge products", e.g. $dx \wedge dy$, can be interpreted as some kind of oriented area elements, $dx \wedge dy = -dy \wedge dx$, etc.)

The exterior derivative of a k-form in $\mathbb{R}^3$ is defined as the $(k+1)$-form from above—and in $\mathbb{R}^n$ if, e.g.,

$$\omega^{(k)} = \sum_{\substack{i_1 < i_2 < \cdots < i_k \\ \forall i_v \in 1,\ldots,n}} a_{i_1,\ldots,i_k}\,dx_{i_1} \wedge \cdots \wedge dx_{i_k},$$

then the exterior derivative d leads to

$$d\omega^{(k)} = \sum_{\substack{j=1 \\ i_1 < \cdots < i_k}}^{n} \frac{\partial a_{i_1,\ldots,i_k}}{\partial x_j}\,dx_j \wedge dx_{i_1} \wedge \cdots \wedge dx_{i_k}.$$

The exterior derivative of a 1-form is therefore a 2-form, and that of a 2-form is a 3-form. On the other hand, because of the interchangeability of mixed derivatives, e.g. because of

$$\frac{\partial^2}{\partial x \partial y} = \frac{\partial^2}{\partial y \partial x},$$

the twofold application of the exterior derivative leads to 0.

Thus, denoting the space of k-forms by $\Omega^k(\mathbb{R}^3)$ and the exterior derivative by d one gets a sequence:

$$0 \xrightarrow{d} \Omega^0(\mathbf{R}^3) \xrightarrow{d} \Omega^1(\mathbf{R}^3) \xrightarrow{d} \Omega^2(\mathbf{R}^3) \xrightarrow{d} \Omega^3(\mathbf{R}^3) \xrightarrow{d} 0.$$

Here $\Omega^k(\mathbb{R}^n)$ is the space of sections of the exterior algebra $\Lambda^k(\mathbb{R}^n)$ vector bundle over $\mathbb{R}^n$, whose dimension is the binomial coefficient $\binom{n}{k}$; note that $\Omega^k(\mathbb{R}^3) = 0$ for $k > 3$ or $k < 0$. Writing only dimensions, one obtains a row of Pascal's triangle:

$$0 \to 1 \to 3 \to 3 \to 1 \to 0;$$

the 1-dimensional fibers correspond to functions, and the 3-dimensional fibers to vector fields, as described below. Modulo suitable identifications, the three nontrivial occurrences of the exterior derivative correspond to grad, curl, and div.

Differential forms and the differential can be defined on any Euclidean space, or indeed any manifold, without any notion of a Riemannian metric. On a Riemannian manifold, or more generally pseudo-Riemannian manifold, k-forms can be identified with k-vector fields (k-forms are k-covector fields, and a pseudo-Riemannian metric gives an isomorphism between vectors and covectors), and on an *oriented* vector space with a nondegenerate form (an isomorphism between vectors and covectors), there is an isomorphism between k-vectors and $(n - k)$-vectors; in particular on (the tangent space of) an oriented pseudo-Riemannian manifold. Thus on an oriented pseudo-Riemannian manifold, one can interchange k-forms, k-vector fields, $(n - k)$-forms, and $(n - k)$-vector fields; this is known as Hodge duality. Concretely, on $\mathbb{R}^3$ this is given by:

- 1-forms and 1-vector fields: the 1-form $a_x\, dx + a_y\, dy + a_z\, dz$ corresponds to the vector field (a_x, a_y, a_z).
- 1-forms and 2-forms: one replaces dx by the dual quantity $dy \wedge dz$ (i.e., omit dx), and likewise, taking care of orientation: dy corresponds to $dz \wedge dx = -dx \wedge dz$, and dz corresponds to $dx \wedge dy$. Thus the form $a_x\, dx + a_y\, dy + a_z\, dz$ corresponds to the "dual form" $a_z\, dx \wedge dy + a_y\, dz \wedge dx + a_x\, dy \wedge dz$.

Thus, identifying 0-forms and 3-forms with functions, and 1-forms and 2-forms with vector fields:

- grad takes a function (0-form) to a vector field (1-form);
- curl takes a vector field (1-form) to a vector field (2-form);
- div takes a vector field (2-form) to a function (3-form)

On the other hand, the fact that $d^2 = 0$ corresponds to the identities

$$\nabla \times (\nabla f) = 0$$

for any function *f*, and

$$\nabla \cdot (\nabla \times \mathbf{v}) = 0$$

for any vector field **v**.

Grad and div generalize to all oriented pseudo-Riemannian manifolds, with the same geometric interpretation, because the spaces of 0-forms and *n*-forms is always (fiberwise) 1-dimensional and can be identified with scalar functions, while the spaces of 1-forms and (n – 1)-forms are always fiberwise *n*-dimensional and can be identified with vector fields.

Curl does not generalize in this way to 4 or more dimensions (or down to 2 or fewer dimensions); in 4 dimensions the dimensions are

$$0 \to 1 \to 4 \to 6 \to 4 \to 1 \to 0;$$

so the curl of a 1-vector field (fiberwise 4-dimensional) is a *2-vector field*, which is fiberwise 6-dimensional, one has

$$\omega^{(2)} = \sum_{i<k=1,2,3,4} a_{i,k}\, dx_i \wedge dx_k,$$

which yields a sum of six independent terms, and cannot be identified with a 1-vector field. Nor can one meaningfully go from a 1-vector field to a 2-vector field to a 3-vector field (4 → 6 → 4), as taking the differential twice yields zero ($d^2 = 0$). Thus there is no curl function from vector fields to vector fields in other dimensions arising in this way.

However, one can define a curl of a vector field as a *2-vector field* in general, as described below.

Curl Geometrically

2-vectors correspond to the exterior power $\Lambda^2 V$; in the presence of an inner product, in coordinates these are the skew-ymmetric matrices, which are geometrically considered as the special orthogonal Lie algebra $\mathfrak{so}$ (V) of infinitesimal rotations. This has $\binom{n}{2} = \frac{1}{2}n(n-1)$ dimensions, and allows one to interpret the differential of a 1-vector fieldas its infinitesimal rotations. Only in 3 dimensions (or trivially in 0 dimensions) does $n = \frac{1}{2}n(n-1)$, which is the most elegant and common case. In 2 dimensions the curl of a vector field is not a vector field but a function, as 2-dimensional rotations are given by an angle (a scalar – an orientation is required to choose whether one counts clockwise or counterclockwise rotations as positive); this is not the div, but is rather perpendicular to it. In 3 dimensions the curl of a vector field is a vector field as is familiar (in 1 and 0 dimensions the curl of a vector field is 0, because there are no non-trivial 2-vectors), while in 4 dimensions the curl of a vector field is, geometrically, at each point an element of the 6-dimensional Lie algebra $\mathfrak{so}$.

Note also that the curl of a 3-dimensional vector field which only depends on 2 coordinates (say x and y) is simply a vertical vector field (in the z direction) whose magnitude is the curl of the 2-dimensional vector field.

Considering curl as a 2-vector field (an antisymmetric 2-tensor) has been used to generalize vector calculus and associated physics to higher dimensions.

Taking the Velocity Field, v(x,y) , as a Vivid Example for the Geometrical Explanation of Curl

In order to explain curl v in a manner as simple as possible we imagine the vectors of a "whirl-field", which may exist in a fluid, to be parallel to (or to lie directly in) the x,y – plane: v = v(x,y) = (v_x,v_y). And we will stick to this assumption in all our ad-hoc explanations below. What we want to investigate geometrically is the relation

$v = k \cdot r^n$ (1- a) (vector v being perpendicular to the radius vector r),

which is representing a smoothly increasing (n > 0) or decreasing (n < 0) function "v over r", and which is suitable for discussing those (few) cases which indeed can be met in (engineering) practice. k is a constant of arbitrary value, i.e. we can also write k = 1, which however does not mean that k is left without dimension (!): k [$1/(\sec m^{n-1})$]. Now, "*rot*" is quite generally referred to the rotary motion of an infinitesimally small fluid-element or liquid particle (in short L.P.). In general this motion, i.e. that of an individual particle of the said kind, has 2 components, which remain unchanged here over time:

(a) The 'revolution' around the centre of the (macroscopically extended) whirl, which we locate in the origin of our coordinate system, x = y = 0 = "point zero", so-to-say. This our system, " Σ(x,y) " by designation, is supposed to be at rest relative to the ground; i.e. the circular 'basin' included, within which the (incompressible) fluid moves on circles always parallel to the earth's surface.

(b) The rotation of an envisaged small L.P. (at a certain distance, r, from the centre of the whirl) around its body axis: A 'microscopic' rotation which occurs additionally to its 'macroscopic' revolution around "point zero". While the latter motion is governed by the angular velocity ω_{mac} (r) = v(r) /r, the former one is characterised by ω_{mic} (r). As we will see later on, it depends on n in eq. (1- a) whether ω_{mic} does exist as a separately distinguishable rotational component or not. What should be stressed at last, is that we are establishing here an idealised, model-like experimental arrangement or – better to say – "thought experiment":

Firstly it is assumed that the fluid in the circular basin or trough has a very low viscosity, thus leading to a rather constant circular motion of the fluid over a longer time of observation.

Secondly we replace an unobservable, tiny liquid particle by a visible (though equally small) solid object in the form of a dark coated Paddle Wheel (in short: P.W.), thus enabling us to see any kind of 'elementary' circular motion, be it a macro- or microscopic one. In order to ensure that the P.W. always moves exactly on a circle around the zero point (x=y=0) its axis is pivoted in an arm (of an effective length **r**), which in turn is pivoted at the zero point. So that arm may also be moved manually in addition to its motion, which is normally effected only by the floating P.W.

Thirdly we want to define a *standard situation or case* as a 'point of reference' for certain other cases concerning the P.W.: In the standard case the P.W. is freely floating with the fluid, i.e. without any surplus motion along its circle around the zero point in Σ(x,y). Hence (in the standard or "normal" case) there is no separate or additional (manual) action on the connecting arm (or rod) itself.

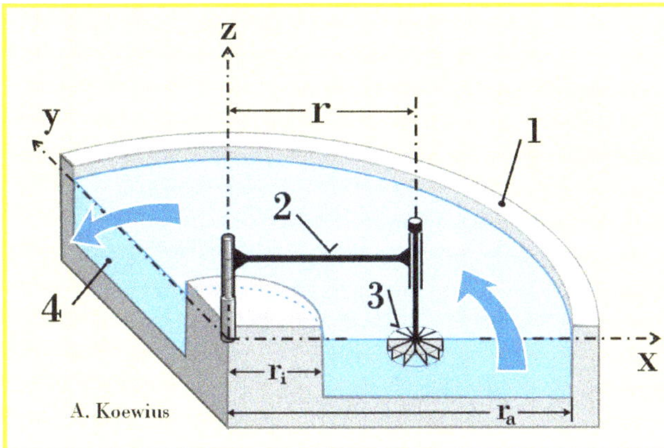

Experimental device suited for the determination of "curl" in a fluid, which is subject to a rotary movement in a circular trough.

1 = section through the circular trough which is fixed on the ground

2 = connecting arm/rod, pivoted in the centre of the whole device and carrying the Paddle Wheel (3) at its free end. Both 2 and 3 are assumed to be pivoted without any friction.

3 = Paddle Wheel (P.W.), pivoted at the end of the arm (2). The arm is carried along by the P.W. The 'infinitesimally small' P.W. had to be drawn here by far larger with good reason.

4 = a fluid which is circulating in the trough (1) under certain boundary conditions and assumptions concerning v(r), i.e. the velocity distribution over r in the trough between r_a and r_i.

Some Definitions Involving Div, Curl and Grad

- A vector field with zero divergence is said to be solenoidal.
- A vector field with zero curl is said to be irrotational.
- A scalar field with zero gradient is said to be, er, constant.

Gradient Theorem

The gradient theorem, also known as the fundamental theorem of line integrals, is a theorem which states that a line integral taken over a vector field which is the gradient of a scalar function can be evaluated only by looking at the endpoints of the scalar function. In mathematical terms,

$$\int_{p1}^{p2} \nabla f \cdot d\vec{r} = f(p_2) - f(p_1)$$

As a corollary, the line integral over ∇f is path independent, therefore any closed path over ∇f will be equal to zero. Gradient vector fields are also known as conservative.

Proof

Let f be a differentiable function and ∇f be its gradient.

$$\int_\gamma \nabla f(u)du = \int_a^b \nabla f(r(t)\cdot r'(t)dt$$

Since the derivative of f with respect to t will be

$$\frac{d}{dt}(f(r(t)) = \nabla f(r(t))\cdot r'(t)$$

by the multivariable chain rule, this expression becomes

$$\int_b^a \frac{d}{dt}(f(r(t))dt = f(r(b)) - f(r(a))$$

by the fundamental theorem of calculus.

Example

Suppose $\gamma \subset \mathbb{R}^2$ is the circular arc oriented counterclockwise from (5, 0) to (−4, 3). Using the definition of a line integral,

$$\begin{aligned}
\int_\gamma y\mathrm{d}x + x\mathrm{d}y &= \int_0^{\pi-\tan^{-1}\left(\frac{3}{4}\right)} ((5\sin t)(-5\sin t) + (5\cos t)(5\cos t))\mathrm{d}t \\
&= \int_0^{\pi-\tan^{-1}\left(\frac{3}{4}\right)} 25\left(-\sin^2 t + \cos^2 t\right)\mathrm{d}t \\
&= \int_0^{\pi-\tan^{-1}\left(\frac{3}{4}\right)} 25\cos(2t)\,\mathrm{d}t \\
&= \frac{25}{2}\sin(2t)\Big|_0^{\pi-\tan^{-1}\left(\frac{3}{4}\right)} \\
&= \frac{25}{2}\sin\left(2\pi - 2\tan^{-1}\left(\frac{3}{4}\right)\right) \\
&= -\frac{25}{2}\sin\left(2\tan^{-1}\left(\frac{3}{4}\right)\right) \\
&= -\frac{25\left(\frac{3}{4}\right)}{\left(\frac{3}{4}\right)^2+1} = -12.
\end{aligned}$$

Notice all of the painstaking computations involved in directly calculating the integral. Instead, since the function $f(x, y) = xy$ is differentiable on all of $\mathbb{R}^2$, we can simply use the gradient theorem to say

$$\int_\gamma y\,dx + x\,dy = \int_\gamma \nabla(xy)\cdot(dx,dy) = xy\,\Big|_{(5,0)}^{(-4,3)} = -4\cdot 3 - 5\cdot 0 = -12.$$

Notice that either way gives the same answer, but using the latter method, most of the work is already done in the proof of the gradient theorem.

Example

For a more abstract example, suppose $\gamma \subset \mathbb{R}^n$ has endpoints p, q, with orientation from p to q. For u in $\mathbb{R}^n$, let |u| denote the Euclidean norm of u. If $\alpha \geq 1$ is a real number, then

$$\int_\gamma |\mathbf{x}|^{\alpha-1}\,\mathbf{x}\cdot d\mathbf{x} = \frac{1}{\alpha+1}\int_\gamma (\alpha+1)\,|\mathbf{x}|^{(\alpha+1)-2}\,\mathbf{x}\cdot d\mathbf{x}$$

$$= \frac{1}{\alpha+1}\int_\gamma \nabla|\mathbf{x}|^{\alpha+1}\cdot d\mathbf{x} = \frac{|\mathbf{q}|^{\alpha+1} - |\mathbf{p}|^{\alpha+1}}{\alpha+1}$$

Here the final equality follows by the gradient theorem, since the function $f(\mathbf{x}) = |\mathbf{x}|^{\alpha+1}$ is differentiable on $\mathbb{R}^n$ if $\alpha \geq 1$.

If $\alpha < 1$ then this equality will still hold in most cases, but caution must be taken if γ passes through or encloses the origin, because the integrand vector field $|x|^{\alpha-1}x$ will fail to be defined there. However, the case $\alpha = -1$ is somewhat different; in this case, the integrand becomes $|x|^{-2}x = \nabla(\log|\mathbf{x}|)$, so that the final equality becomes $\log|q| - \log|p|$.

Note that if $n = 1$, then this example is simply a slight variant of the familiar power rule from single-variable calculus.

Suppose there are n point charges arranged in three-dimensional space, and the i-th point charge has charge Q_i and is located at position p_i in $\mathbb{R}^3$. We would like to calculate the work done on a particle of charge q as it travels from a point a to a point b in $\mathbb{R}^3$. Using Coulomb's law, we can easily determine that the force on the particle at position r will be

$$\mathbf{F}(\mathbf{r}) = kq\sum_{i=1}^{n}\frac{Q_i(\mathbf{r}-\mathbf{p}_i)}{|\mathbf{r}-\mathbf{p}_i|^3}$$

Here |u| denotes the Euclidean norm of the vector u in $\mathbb{R}^3$, and $k = 1/(4\pi\varepsilon_0)$, where ε_0 is the vacuum permittivity.

Let $\gamma \subset \mathbb{R}^3 - \{p_1, ..., p_n\}$ be an arbitrary differentiable curve from a to b. Then the work done on the particle is

$$W = \int_\gamma \mathbf{F}(\mathbf{r})\cdot d\mathbf{r} = \int_\gamma\left(kq\sum_{i=1}^{n}\frac{Q_i(\mathbf{r}-\mathbf{p}_i)}{|\mathbf{r}-\mathbf{p}_i|^3}\right)\cdot d\mathbf{r} = kq\sum_{i=1}^{n}\left(Q_i\int_\gamma\frac{\mathbf{r}-\mathbf{p}_i}{|\mathbf{r}-\mathbf{p}_i|^3}\cdot d\mathbf{r}\right)$$

Now for each i, direct computation shows that

$$\frac{\mathbf{r}-\mathbf{p}_i}{|\mathbf{r}-\mathbf{p}_i|^3} = -\nabla\frac{1}{|\mathbf{r}-\mathbf{p}_i|}.$$

Thus, continuing from above and using the gradient theorem,

$$W = -kq\sum_{i=1}^{n}\left(Q_i\int_\gamma \nabla\frac{1}{|\mathbf{r}-\mathbf{p}_i|}\cdot d\mathbf{r}\right) = kq\sum_{i=1}^{n}Q_i\left(\frac{1}{|\mathbf{a}-\mathbf{p}_i|}-\frac{1}{|\mathbf{b}-\mathbf{p}_i|}\right)$$

We are finished. Of course, we could have easily completed this calculation using the powerful language of electrostatic potential or electrostatic potential energy (with the familiar formulas $W = -\Delta U = -q\Delta V$). However, we have not yet *defined* potential or potential energy, because the *converse* of the gradient theorem is required to prove that these are well-defined, differentiable functions and that these formulas hold. Thus, we have solved this problem using only Coulomb's Law, the definition of work, and the gradient theorem.

Converse of the Gradient Theorem

The gradient theorem states that if the vector field F is the gradient of some scalar-valued function (i.e., if F is conservative), then F is a path-independent vector field (i.e., the integral of F over some piecewise-differentiable curve is dependent only on end points). This theorem has a powerful converse:

If F is a path-independent vector field, then F is the gradient of some scalar-valued function.

It is straightforward to show that a vector field is path-independent if and only if the integral of the vector field over every closed loop in its domain is zero. Thus the converse can alternatively be stated as follows: If the integral of F over every closed loop in the domain of F is zero, then F is the gradient of some scalar-valued function.

Proof of the Converse

Suppose U is an open, path-connected subset of $\mathbb{R}^n$, and $F : U \to \mathbb{R}^n$ is a continuous and path-independent vector field. Fix some element a of U, and define $f : U \to \mathbb{R}$ by

$$f(\mathbf{x}) := \int_{\gamma[\mathbf{a},\mathbf{x}]} \mathbf{F}(\mathbf{u})\cdot d\mathbf{u}$$

Here $\gamma[\mathbf{a}, \mathbf{x}]$ is any (differentiable) curve in U originating at **a** and terminating at **x**. We know that f is well-defined because **F** is path-independent.

Let **v** be any nonzero vector in $\mathbb{R}^n$. By the definition of the directional derivative,

$$\begin{aligned}\frac{\partial f}{\partial \mathbf{v}}(\mathbf{x}) &= \lim_{t\to 0}\frac{f(\mathbf{x}+t\mathbf{v})-f(\mathbf{x})}{t}\\ &= \lim_{t\to 0}\frac{\int_{\gamma[\mathbf{a},\mathbf{x}+t\mathbf{v}]}\mathbf{F}(\mathbf{u})\cdot d\mathbf{u}-\int_{\gamma[\mathbf{a},\mathbf{x}]}\mathbf{F}(\mathbf{u})\cdot d\mathbf{u}}{t}\end{aligned}$$

$$= \lim_{t\to 0}\frac{1}{t}\int_{\gamma[\mathbf{x},\mathbf{x}+t\mathbf{v}]}\mathbf{F}(\mathbf{u})\cdot d\mathbf{u}$$

To calculate the integral within the final limit, we must parametrize $\gamma[x, x + tv]$. Since F is path-independent, U is open, and t is approaching zero, we may assume that this path is a straight line, and parametrize it as $\mathbf{u}(s) = \mathbf{x} + s\mathbf{v}$ for $0 < s < t$. Now, since $\mathbf{u}'(s) = \mathbf{v}$, the limit becomes

$$\lim_{t\to 0}\frac{1}{t}\int_0^t \mathbf{F}(\mathbf{u}(s))\cdot\mathbf{u}'(s)ds = \frac{d}{dt}\int_0^t \mathbf{F}(\mathbf{x}+s\mathbf{v})\cdot\mathbf{v}ds\,\Big|_{t=0} = \mathbf{F}(\mathbf{x})\cdot\mathbf{v}$$

Thus we have a formula for $\partial_{\mathbf{v}} f$, where $\mathbf{v}$ is arbitrary. Let $\mathbf{x} = (x_1, x_2, \ldots, x_n)$ and let $\mathbf{e}_i$ denote the i-th standard basis vector, so that

$$\nabla f(\mathbf{x}) = \left(\frac{\partial f(\mathbf{x})}{\partial x_1}, \frac{\partial f(\mathbf{x})}{\partial x_2}, \ldots, \frac{\partial f(\mathbf{x})}{\partial x_n}\right) = (\mathbf{F}(\mathbf{x})\cdot\mathbf{e}_1, \mathbf{F}(\mathbf{x})\cdot\mathbf{e}_2, \ldots, \mathbf{F}(\mathbf{x})\cdot\mathbf{e}_n) = \mathbf{F}(\mathbf{x})$$

Thus we have found a scalar-valued function f whose gradient is the path-independent vector field F, as desired.

Example of the Converse Principle

To illustrate the power of this converse principle, we cite an example that has significant physical consequences. In classical electromagnetism, the electric force is a path-independent force; i.e. the work done on a particle that has returned to its original position within an electric field is zero (assuming that no changing magnetic fields are present).

Therefore, the above theorem implies that the electric force field $F_e : S \to \mathbb{R}^3$ is conservative (here S is some open, path-connected subset of $\mathbb{R}^3$ that contains a charge distribution). Following the ideas of the above proof, we can set some reference point a in S, and define a function $U_e : S \to \mathbb{R}$ by

$$U_e(\mathbf{r}) := -\int_{\gamma[\mathbf{a},\mathbf{r}]}\mathbf{F}_e(\mathbf{u})\cdot d\mathbf{u}$$

Using the above proof, we know U_e is well-defined and differentiable, and $\mathbf{F}_e = -\nabla U_e$ (from this formula we can use the gradient theorem to easily derive the well-known formula for calculating work done by conservative forces: $W = -\Delta U$). This function U_e is often referred to as the electrostatic potential energy of the system of charges in S (with reference to the zero-of-potential **a**). In many cases, the domain S is assumed to be unbounded and the reference point **a** is taken to be "infinity", which can be made rigorous using limiting techniques. This function U_e is an indispensable tool used in the analysis of many physical systems.

Generalizations

Many of the critical theorems of vector calculus generalize elegantly to statements about the integration of differential forms on manifolds. In the language of differential forms and exterior derivatives, the gradient theorem states that

$$\int_{\partial\gamma} \phi = \int_{\gamma} d\phi$$

for any 0-form, ϕ, defined on some differentiable curve $\gamma \subset \mathbb{R}^n$ (here the integral of ϕ over the boundary of the γ is understood to be the evaluation of at the endpoints of γ).

Notice the striking similarity between this statement and the generalized version of Stokes' theorem, which says that the integral of any compactly supported differential form ω over the boundary of some orientable manifold Ω is equal to the integral of its exterior derivative $d\omega$ over the whole of Ω, i.e.,

$$\int_{\partial\Omega} \omega = \int_{\Omega} d\omega$$

This powerful statement is a generalization of the gradient theorem from 1-forms defined on one-dimensional manifolds to differential forms defined on manifolds of arbitrary dimension.

The converse statement of the gradient theorem also has a powerful generalization in terms of differential forms on manifolds. In particular, suppose ω is a form defined on a contractible domain, and the integral of ω over any closed manifold is zero. Then there exists a form ψ such that $\omega = d\psi$. Thus, on a contractible domain, every closed form is exact. This result is summarized by the Poincaré lemma.

Divergence Theorem

The divergence theorem (also called Gauss's theorem or Gauss-Ostrogradsky theorem) is a theorem which relates the flux of a vector field through a closed surface to the vector field inside the surface. The theorem states that the outward flux of a vector field through a closed surface is equal to the triple integral of the divergence of the vector field inside the surface.

What this theorem actually states is the physical fact that in the absence of the creation or destruction of matter, the density within a region of space can change only by having it flow into or away from the region through its boundary.

The theorem is very applicable in different areas of physics, among others electrostatics and fluid dynamics. Another very important application of the theorem is that several physical laws can be written in both differential and integral form.

Statement: "The volume integral of the divergence of a vector field A taken over any volume Vbounded by a closed surfaceS is equal to the surface integral of A over the surfaceS."

Or

"The volume integral of the divergence of a vector field over the volume enclosed by surface S isequal to the flux of that vector field taken over that surface S."

Hence, this theorem is used to convert volume integral into surface integral.

Under suitable conditions, if E is a region of three dimensional space and D is its boundary surface, oriented outward, then

$$\iint_D \mathbf{F}\cdot\mathbf{N}dS = \iiint_E \nabla\cdot\mathbf{F}dV.$$

Proof: Again this theorem is too difficult to prove here, but a special case is easier. In the proof of a special case of Green's Theorem, we needed to know that we could describe the region of integration in both possible orders, so that we could set up one double integral using *dxdy* and another using *dydx* . Similarly here, we need to be able to describe the three-dimensional region *E* in different ways.

We start by rewriting the triple integral:

$$\iiint_E \nabla\cdot\mathbf{F}dV = \iiint_E (P_x + Q_y + R_z)dV = \iiint_E P_x dV + \iiint_E Q_y dV + \iiint_E R_z dV.$$

The double integral may be rewritten:

$$\iint_D F\cdot \mathrm{N}dS = \iint_D (Pi + Qj + Rk)\cdot \mathrm{N}dS = \iint_D Pi\cdot \mathrm{N}dS + \iint_D Qj\cdot \mathrm{N}dS + \iint_D Rk\cdot \mathrm{N}dS.$$

To prove that these give the same value it is sufficient to prove that

$$\iint_D Pi\cdot \mathrm{N}\,dS = \iiint_E P_x dV$$

$$\iint_D Qj\cdot \mathrm{N}\,dS = \iiint_E Q_y dV \quad ,and$$

$$\iint_D Rk\cdot \mathrm{N}\,dS = \iiint_E R_z dV.$$

We set the triple integral up with *dx* dx innermost:

$$\iiint_E P_x dV = \iint_B \int_{g_1(y,z)}^{g_2(y,z)} P_x dxdA = \iint_B P(g_2(y,z),y,z) - P(g_1(y,z),y,z)dA,$$

where *B* is the region in the *y* -*z* plane over which we integrate. The boundary surface of *E* consists of a "top" $x = g_2(y,z)$, a "bottom" $x = g_1(y,z)$, and a "wrap-around side" that is vertical to the *y* -*z* plane. To integrate over the entire boundary surface, we can integrate over each of these (top, bottom, side) and add the results. Over the side surface, the vector N is perpendicular to the vector i, so

$$\iint_{\text{side}} P\mathbf{i}\cdot\mathbf{N}dS = \iint_{\text{side}} 0dS = 0.$$

Thus, we are left with just the surface integral over the top plus the surface integral over the bottom.

For the top, we use the vector function $\mathbf{r} = \langle g_2(y,z), y, z \rangle$ which gives $\mathbf{r}_y \times \mathbf{r}_z = \langle 1, -g_{2y}, -g_{2z} \rangle$; the dot product of this with $\mathbf{i} = \langle 1,0,0 \rangle$ is 1. Then

$$\iint_{\text{top}} P\mathbf{i} \cdot \mathbf{N} dS = \iint_B P(g_2(y,z), y, z) dA.$$

In almost identical fashion we get

$$\iint_{\text{bottom}} P\mathbf{i} \cdot \mathbf{N} dS = -\iint_B P(g_1(y,z), y, z) dA,$$

where the negative sign is needed to make **N** point in the negative x direction. Now

$$\iint_D P\mathbf{i} \cdot \mathbf{N} dS = \iint_B P(g_2(y,z), y, z) dA - \iint_B P(g_1(y,z), y, z) dA,$$

which is the same as the value of the triple integral above.

Example: Let $F = \langle 2x, 3y, z^2 \rangle$, and consider the three-dimensional volume inside the cube with faces parallel to the principal planes and opposite corners at $(0,0,0)$ and $(1,1,1)$. We compute the two integrals of the divergence theorem.

The triple integral is the easier of the two:

$$\int_0^1 \int_0^1 \int_0^1 2 + 3 + 2z \, dx \, dy \, dz = 6.$$

The surface integral must be separated into six parts, one for each face of the cube. One face is $z=0$ or $\mathbf{r}=\langle u,v,0 \rangle$, $0 \le u,v \le 1$. Then $\mathbf{r}_u = \langle 1,0,0 \rangle$, $\mathbf{r}_v = \langle 0,1,0 \rangle$, and $\mathbf{r}_u \times \mathbf{r}_v = \langle 0,0,1 \rangle$. We need this to be oriented downward (out of the cube), so we use $\langle 0,0,-1 \rangle$ and the corresponding integral is

$$\int_0^1 \int_0^1 -z^2 \, du \, dv = \int_0^1 \int_0^1 0 \, du \, dv = 0.$$

Another face is $y=1$ or $\mathbf{r}=\langle u,1,v \rangle$. Then $\mathbf{r}_u = \langle 1,0,0 \rangle$, $\mathbf{r}_v = \langle 0,0,1 \rangle$, and $\mathbf{r}_u \times \mathbf{r}_v = \langle 0,-1,0 \rangle$. We need a normal in the positive y direction, so we convert this to $\langle 0,1,0 \rangle$, and the corresponding integral is

$$\int_0^1 \int_0^1 3y \, du \, dv = \int_0^1 \int_0^1 3 \, du \, dv = 3.$$

The remaining four integrals have values 0, 0, 2, and 1, and the sum of these is 6, in agreement with the triple integral.

Example: Let $F=\langle x^3, y^3, z^2 \rangle$, and consider the cylindrical volume $x^2 + y^2 \le 9$, $0 \le z \le 2$. The triple integral (using cylindrical coordinates) is

$$\int_0^{2\pi} \int_0^3 \int_0^3 (3r^2 + 2z) r \, dz \, dr \, d\theta = 279\pi.$$

For the surface we need three integrals. The top of the cylinder can be represented by $\mathbf{r}=\langle v\cos u, v\sin u, 2 \rangle$; $\mathbf{r}_u \times \mathbf{r}_v = \langle 0,0,-v \rangle$, which points down into the cylinder, so we convert it to $\langle 0,0,v \rangle$. Then

$$\int_0^{2\pi}\int_0^3 \left\langle v^3\cos^3 u, v^3\sin^3 u, 4\right\rangle\cdot\left\langle 0,0,-v\right\rangle dvdu = \int_0^{2\pi}\int_0^3 4u\; dvdu = 36\pi.$$

The bottom is r=⟨$v$cos u,$v$sin u,0⟩ ; $\mathbf{r}_u \times \mathbf{r}_v$ =⟨0,0,−v⟩ and

$$\int_0^{2\pi}\int_0^3 \left\langle v^3\cos^3 u, v^3\sin^3 u, 0\right\rangle\cdot\left\langle 0,0,-v\right\rangle dvdu = \int_0^{2\pi}\int_0^3 0dvdu = 0.$$

The side of the cylinder is r=⟨3cos u, 3sin u, v⟩; $\mathbf{r}_u \times \mathbf{r}_v$ =⟨3cos u, 3sin u, 0⟩ which does point outward, so

$$\int_0^{2\pi}\int_0^2 \left\langle 27cos^3 u, 27sin^3 u, v^2\right\rangle\cdot\left\langle 3cos\,u, 3sin\,u, 0\right\rangle dv\,du$$

$$\int_0^{2\pi}\int_0^2 81cos^4 u + 81sin^4 u\,dv\,du = 243\pi.$$

The total surface integral is thus $36\pi + 0 + 243\pi = 279\pi$.

Corollaries

By replacing $\mathbf{F}$ in the divergence theorem with specific forms, other useful identities can be derived (cf. vector identities).

- With $\mathbf{F}\rightarrow\mathbf{F}g$ for a scalar function g and a vector field F,

  $$\iiint_V \left[\mathbf{F}\cdot(\nabla g) + g(\nabla\cdot\mathbf{F})\right]dV = \oiint_S g\mathbf{F}\cdot\mathbf{n}dS.$$

  A special case of this is $\mathbf{F} = \nabla f$, in which case the theorem is the basis for Green's identities.

- With $\mathbf{F}\rightarrow\mathbf{F}\times\mathbf{G}$ for two vector fields F and G,

  $$\iiint_V \left[\mathbf{G}\cdot(\nabla\times\mathbf{F}) - \mathbf{F}\cdot(\nabla\times\mathbf{G})\right]dV = \oiint_S (\mathbf{F}\times\mathbf{G})\cdot d\mathbf{S}.$$

- With $\mathbf{F}\rightarrow f\mathbf{c}$ for a scalar function f and vector field c:

  $$\iiint_V \mathbf{c}\cdot\nabla f\,dV = \oiint_S (\mathbf{c}f)\cdot d\mathbf{S} - \iiint_V f(\nabla\cdot\mathbf{c})dV.$$

  The last term on the right vanishes for constant or any divergence free (solenoidal) vector field, e.g. Incompressible flows without sources or sinks such as phase change or chemical reactions etc.

- With $\mathbf{F}\rightarrow\mathbf{c}\times\mathbf{F}$ for vector field F and constant vector c:

  $$\iiint_V \mathbf{c}\cdot(\nabla\times\mathbf{F})dV = \oiint_S (\mathbf{F}\times\mathbf{c})\cdot d\mathbf{S}.$$

Example

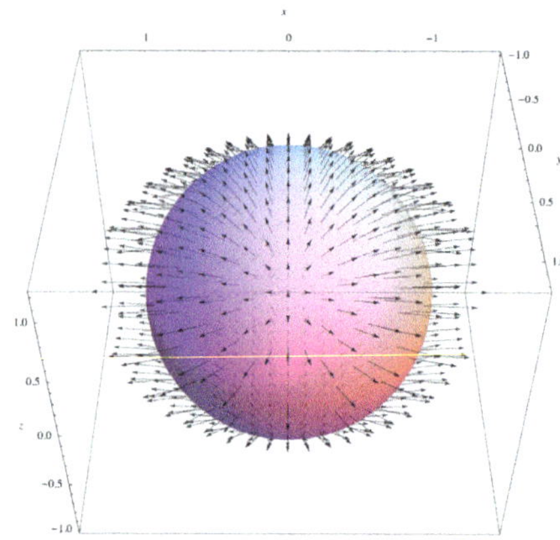

The vector field corresponding to the example shown

Suppose we wish to evaluate

$$\oiint_S \mathbf{F} \cdot \mathbf{n} dS,$$

where S is the unit sphere defined by

$$S = \left\{(x, y, z) \in \mathbb{R}^3 : x^2 + y^2 + z^2 = 1\right\},$$

and $\mathbf{F}$ is the vector field

$$\mathbf{F} = 2x\mathbf{i} + y^2\mathbf{j} + z^2\mathbf{k}.$$

The direct computation of this integral is quite difficult, but we can simplify the derivation of the result using the divergence theorem, because the divergence theorem says that the integral is equal to:

$$\iiint_W (\nabla \cdot \mathbf{F}) dV = 2\iiint_W (1 + y + z) dV = 2\iiint_W dV + 2\iiint_W y dV + 2\iiint_W z dV,$$

where W is the unit ball:

$$W = \left\{(x, y, z) \in \mathbb{R}^3 : x^2 + y^2 + z^2 \leq 1\right\}.$$

Since the function y is positive in one hemisphere of W and negative in the other, in an equal and opposite way, its total integral over W is zero. The same is true for z:

$$\iiint_W y dV = \iiint_W z dV = 0.$$

Therefore,

$$\oiint_S \mathbf{F} \cdot \mathbf{n} dS = 2\iiint_W dV = \frac{8\pi}{3},$$

because the unit ball W has volume $\frac{4\pi}{3}$.

Applications

Differential form and Integral form of Physical Laws

As a result of the divergence theorem, a host of physical laws can be written in both a differential form (where one quantity is the divergence of another) and an integral form (where the flux of one quantity through a closed surface is equal to another quantity). Three examples are Gauss's law (in electrostatics), Gauss's law for magnetism, and Gauss's law for gravity.

Continuity Equations

Continuity equations offer more examples of laws with both differential and integral forms, related to each other by the divergence theorem. In fluid dynamics, electromagnetism, quantum mechanics, relativity theory, and a number of other fields, there are continuity equations that describe the conservation of mass, momentum, energy, probability, or other quantities. Generically, these equations state that the divergence of the flow of the conserved quantity is equal to the distribution of *sources* or *sinks* of that quantity. The divergence theorem states that any such continuity equation can be written in a differential form (in terms of a divergence) and an integral form (in terms of a flux).

Inverse-square Laws

Any *inverse-square law* can instead be written in a *Gauss's law*-type form (with a differential and integral form, as described above). Two examples are Gauss's law (in electrostatics), which follows from the inverse-square Coulomb's law, and Gauss's law for gravity, which follows from the inverse-square Newton's law of universal gravitation. The derivation of the Gauss's law-type equation from the inverse-square formulation or vice versa is exactly the same in both cases.

Examples

To verify the planar variant of the divergence theorem for a region R:

$$R = \left\{(x, y) \in \mathbb{R}^2 : x^2 + y^2 \leq 1\right\},$$

and the vector field:

$$\mathbf{F}(x, y) = 2y\mathbf{i} + 5x\mathbf{j}.$$

The boundary of R is the unit circle, C, that can be represented parametrically by:

$$x = \cos(s), \quad y = \sin(s)$$

such that $0 \leq s \leq 2\pi$ where s units is the length arc from the point $s = 0$ to the point P on C. Then a vector equation of C is

$$C(s) = \cos(s)\mathbf{i} + \sin(s)\mathbf{j}.$$

At a point P on C:

$$P = (\cos(s), \sin(s)) \Rightarrow \mathbf{F} = 2\sin(s)\mathbf{i} + 5\cos(s)\mathbf{j}.$$

Therefore,

$$\oint_C \mathbf{F}\cdot\mathbf{n}ds = \int_0^{2\pi} (2\sin(s)\mathbf{i} + 5\cos(s)\mathbf{j})\cdot(\cos(s)\mathbf{i} + \sin(s)\mathbf{j})ds$$
$$= \int_0^{2\pi} (2\sin(s)\cos(s) + 5\sin(s)\cos(s))ds$$
$$= 7\int_0^{2\pi} \sin(s)\cos(s)ds$$
$$= 0.$$

Because $M = 2y$, $\frac{\partial M}{\partial x} = 0$, and because $N = 5x, \frac{\partial N}{\partial x} = 0$. Thus

$$\iint_R \nabla\cdot\mathbf{F}dA = \iint_R \left(\frac{\partial M}{\partial x} + \frac{\partial N}{\partial y}\right)dA = 0.$$

Generalizations

Multiple Dimensions

One can use the general Stokes' Theorem to equate the n-dimensional volume integral of the divergence of a vector field **F** over a region U to the $(n-1)$-dimensional surface integral of **F** over the boundary of U:

$$\underbrace{\int\cdots\int_U}_{n} \nabla\cdot\ \mathbf{F}\cdot dv = \underbrace{\oint\cdots\oint_{\partial U}}_{n-1} \mathbf{F}\cdot nds$$

This equation is also known as the Divergence theorem.

When $n = 2$, this is equivalent to Green's theorem.

When $n = 1$, it reduces to the Fundamental theorem of calculus.

Tensor Fields

Writing the theorem in Einstein notation:

$$\iiint_V \frac{\partial \mathbf{F}_i}{\partial x_i}dV = \oiint_s \mathbf{F}_i n_i\, dS$$

suggestively, replacing the vector field **F** with a rank-n tensor field T, this can be generalized to:

$$\iiint_V \frac{\partial T_{i_1 i_2\cdots i_q\cdots i_n}}{\partial x_{i_q}}dV = \oiint_S T_{i_1 i_2\cdots i_q\cdots i_n} n_{i_q}\, dS.$$

where on each side, tensor contraction occurs for at least one index. This form of the theorem is still in 3d, each index takes values 1, 2, and 3. It can be generalized further still to higher (or lower) dimensions (for example to 4d spacetime in general relativity).

Stokes' Theorem

Stokes' theorem can be regarded as a higher-dimensional version of Green's Theorem. If we recall from previous lessons, Green's Theorem relates a double integral over a plane region to a line integral around its plane boundary curve. Stoke's Theorem relates a surface integral over a surface to a line integral along the boundary curve.

In fact, Stokes' Theorem provides insight into a physical interpretation of the curl. In a vector field, the rotation of the vector field is at a maximum when the curl of the vector field and the normal vector have the same direction. In other words, while the tendency to rotate will vary from point to point on the surface, Stokes' Theorem says that the collective measure of this rotational tendency taken over the entire surface is equal to the tendency of a fluid to circulate around the boundary curve.

Let S be a surface in space and the boundary of S is simple closed curve c. Let $\vec{F}\ (x, y, z)$ is a continuous function that has continuous partial derivatives in S, then,

$\int_S \left(\vec{\nabla} \times \vec{F}\right) \cdot \hat{n} ds = \oint_C \vec{F} \cdot d\vec{r}$ where $\hat{n}$ is an outward drawn normal to the elemental surface ds and $d\vec{r}$ is taken along C.

Proof of Stokes' Theorem

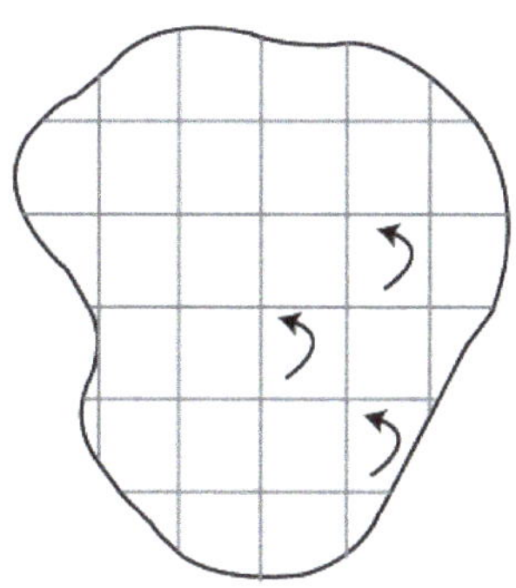

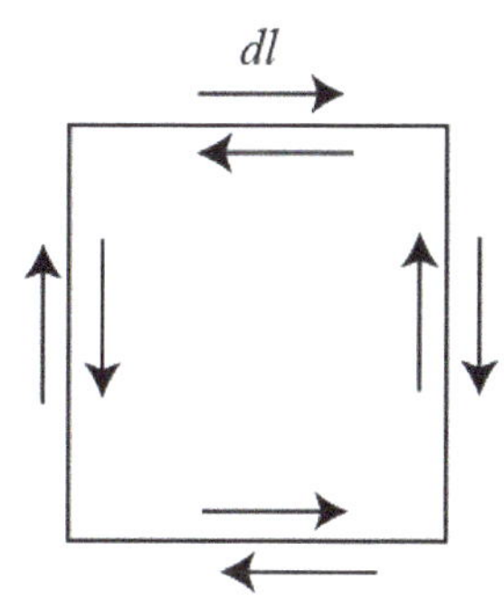

We have shown that circulation around a small mesh is written as,

$$\sum_{4\,sides} \vec{A} . \vec{dl} - \left(\vec{\nabla} \times \vec{A}\right) dxdy$$

The surface integrals (i.e. RHS of the above equation) can be added together. Again (as in the divergence theorem case) the line integrals of the interior line segments cancel identically. Only the integral around the perimeter survives, giving

$$\sum_{external\ line\ segments} \vec{A}.\vec{dl} = \sum_{rectangle}\left(\vec{\nabla} \times \vec{A}\right).d\vec{s}$$

Then, $\oint \vec{A}.\vec{dl} = \int\left(\vec{\nabla} \times \vec{A}\right).d\vec{s}$

Example

Verify Stokes' theorem for $\vec{F}\,(2x - y)\hat{i} - yz^2\hat{j} - y^2 z\hat{k}$ for the paraboloid S devoted by

$$z = f\,(x\,,y) = 1 - (x^2 + y^2) \qquad\qquad z \geq 0$$

Or the upper half surface of a sphere.

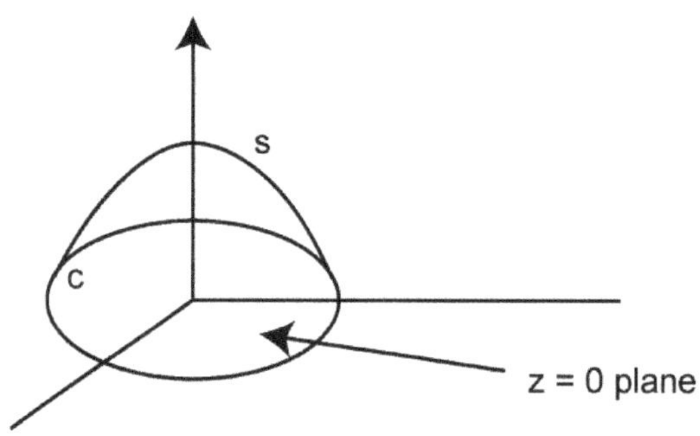

In z = 0 plane the boundary c of the surface S is a circle $(x^2 + y^2) = 1$

A convenient way to determine the line integral (refer to Stoke's theorem) is to substitute

$x = \cos t$, $y = sint$. $0 \leq t \leq 2\pi$ and $z = 0$.

Thus

$$\oint_c \vec{F}.\vec{dr} = \oint_c (2x - y).(dx\,i + dy\,j + dz\,\hat{k})$$

$$= \oint_c (2x - y)dx = \int_0^{2\pi} 2\cos t - \sin t)(-\sin t)\,dt$$

$$= \pi$$

Also

$$\vec{\nabla} \times \vec{F} = \hat{K}(\hat{z})\,, so\left(\vec{\nabla} \times \vec{F}\right).d\vec{s} = \int \hat{z}.\hat{n}\,ds$$

$$\int dxdy = \int_{x=-1}^{1} \int_{y=-\sqrt{1-x^2}}^{\sqrt{1-x^2}} dxdy$$

$= \pi$. (verified)

Examples

Some example based on Stokes' theorem are given below:

Example 1: If F(x, y, z) = xi + zj + 2yk and curve C is obtained by the intersection of circle $x^2 + y^2 = 1$ by the plane z = x, oriented upward. Verify Stokes' theorem.

Solution: Given that $x^2 + y^2 = 1$ and plane z = x

$$r(t) = < cos\ t,\ sin\ t,\ cos\ t >$$

$$dr = < -sin\ t,\ cos\ t,\ -sin\ t > dt$$

F(x, y, z) = xi + zj + 2yk

$$F(t) = < cos\ t,\ cos\ t,\ 2\ sin\ t >$$

F . dr = (-sin t cos t + $\cos^2$t - 2 $\sin^2$ t) dt

$$\int_C F.dr = \int_0^{2\pi} (-sintcost + cos^2t - 2sin^2t)dt$$

$$= \int_0^{2\pi} \left(\frac{-1}{2} \sin 2t + 1 - 3\sin^2 t \right) dt$$

$$= \int_0^{2\pi} \left(\frac{-1}{2} \sin 2t + 1 - \frac{3}{2}(1 - \cos 2t \right) dt$$

$$= \int_0^{2\pi} \left(\frac{-1}{2} \sin\ 2t + \frac{3}{2} \cos\ 2t - \frac{1}{2} \right) dt$$

$$= \left[\frac{1}{4} \cos\ 2t + \frac{3}{4} \sin 2t - \frac{1}{2} t \right]$$

$$= \left(\frac{1}{4} \cos\ 4\pi + \frac{3}{4} \sin\ 4\pi - \pi \right) - \left(\frac{1}{4} \right)$$

$$= -\pi$$

$$F = < x,\ z,\ 2y >$$

$$curl\ F = < 1,\ 0,\ 0 >$$

$$dS = < -1,\ 0,\ 1 > dx\ dy$$

$$\iint_C curl\, F.dS = \iint_{unit\ disc} <1,0,0>.<-1,0,1>\, dx\, dy$$

$$= -\iint_{unit\ disc} dx\, dy$$

$$= -\pi$$

It verifies that $\int_C F.dr = \int_C curl\, F.dS$ which means Stokes' theorem holds.

Example2: Calculate the line integral given by $\int_C F.r(s)ds$, if $F = yi + xz^3 j - zy^3 k$ and curve C is given by the circle $x^2 + y^2 = 4$ and $z = -3$.

Solution: In Stokes' theorem, the normal vector n points towards the direction of positive z; therefore n = k

We are going to evaluate curl $F = \nabla \times F$

$$\begin{vmatrix} i & j & k \\ \dfrac{\partial}{\partial x} & \dfrac{\partial}{\partial y} & \dfrac{\partial}{\partial z} \\ y & xz^3 & -zy^3 \end{vmatrix}$$

$$= -3z(y^2 + xz)i + (z^3 - 1)k$$

$$=< -3z(y^2 + xz), 0, (z^3 - 1) >$$

$$curl\, F.\, n =< -3z(y^2 + xz), 0, (z^3 - 1) >.< 0,0,1 > \text{ (since n=k)}$$

$$= z^3 - 1$$

$$[curl\, F.\, n]_{z=-3} = -27 - 1 = -28$$

$$\iint_{x^2+y^2=4, z=-3} curl\, F.\, n = -28\, dx\, dy$$

$$= -28(2^2)\pi$$

$$= -112\pi$$

Formulation for Smooth Manifolds with Boundary

Let Ω be an oriented smooth manifold with boundary of dimension n and let α be a smooth n-differential form that is compactly supported on Ω. First, suppose that α is compactly supported in the domain of a single, oriented coordinate chart $\{U, \varphi\}$. In this case, we define the integral of α over Ω as

$$\int_\Omega \alpha = \int_{\varphi(U)} \left(\varphi^{-1}\right)^* \alpha,$$

i.e., via the pullback of α to R^n.

More generally, the integral of α over Ω is defined as follows: Let $\{\psi_i\}$ be a partition of unity associated with a locally finitecover $\{U_i, \varphi_i\}$ of (consistently oriented) coordinate charts, then define the integral

$$\int_\Omega \alpha \equiv \sum_i \int_{U_i} \psi_i \alpha,$$

where each term in the sum is evaluated by pulling back to R^n as described above. This quantity is well-defined; that is, it does not depend on the choice of the coordinate charts, nor the partition of unity.

The generalized Stokes' theorem reads:

Theorem: *(Stokes–Cartan) If ω is a smooth $(n-1)$ -form with compact support on smooth n -dimensional manifold-with-boundary Ω, $\partial\Omega$, denotes the boundary of Ω given the induced orientation, and $i:\partial\Omega \hookrightarrow \Omega$ is the inclusion map, then*

$$\int_\Omega d\omega = \int_{\partial\Omega} i^*\omega.$$

Conventionally, $\int_{\partial\Omega} i^*\omega$ is abbreviated as $\int_{\partial\Omega}\omega$,, since the pullback of a differential form by the inclusion map is simply its restriction to its domain: $i^*\omega = \omega|_{\partial\Omega}$. Here d is the exterior derivative, which is defined using the manifold structure only. The right-hand side is sometimes written as $\oint_{\partial\Omega}\omega$ to stress the fact that the $(n-1)-$ manifold $\partial\Omega$ has no boundary. (This fact is also an implication of Stokes' theorem, since for a given smooth n -dimensional manifold Ω,, application of the theorem twice gives $\int_{\partial(\partial\Omega)}\omega = \int_\Omega d(d\omega) = 0$ for any $(n-2)$ -form ω, which implies that $\partial(\partial\Omega) = \varnothing$.) The right-hand side of the equation is often used to formulate *integral* laws; the left-hand side then leads to equivalent *differential* formulations.

The theorem is often used in situations where Ω, is an embedded oriented submanifold of some bigger manifold, often $\mathbf{R}^k$, on which the form ω is defined.

Topological Preliminaries; Integration over Chains

Let M be a smooth manifold. A (smooth) singular k-simplex in M is defined as a smooth map from the standard simplex in R^k to M. The group $C_k(M, Z)$ of singular k-chains on M is defined to be the free abelian group on the set of singular k-simplices in M. These groups, together with the boundary map, ∂, define a chain complex. The corresponding homology (resp. cohomology) group is isomorphic to the usual singular homology group $H_k(M, Z)$ (resp. the singular cohomology group $H^k(M, \mathbf{Z})$), defined using continuous rather than smooth simplices in M.

On the other hand, the differential forms, with exterior derivative, d, as the connecting map, form a cochain complex, which defines the de Rham cohomology groups $H^k{}_{dR}(M, R)$.

Differential k-forms can be integrated over a k-simplex in a natural way, by pulling back to R^k. Extending by linearity allows onc to integrate over chains. This gives a linear map from the space of k-forms to the kth group of singular cochains, $C^k(M, Z)$, the linear functionals on $C^k(M, Z)$. In other words, a k-form ω defines a functional

$$I(\omega)(c) = \oint_c \omega$$

on the k-chains. Stokes' theorem says that this is a chain map from de Rham cohomology to singular cohomology with real coefficients; the exterior derivative, d, behaves like the *dual* of ∂ on forms. This gives a homomorphism from de Rham cohomology to singular cohomology. On the level of forms, this means:

1. closed forms, i.e., $d\omega = 0$, have zero integral over *boundaries*, i.e. over manifolds that can be written as $\partial \Sigma_c M_c$, and
2. exact forms, i.e., $\omega = d\sigma$, have zero integral over *cycles*, i.e. if the boundaries sum up to the empty set: $\Sigma_c M_c = \varnothing$.

De Rham's theorem shows that this homomorphism is in fact an isomorphism. So the converse to 1 and 2 above hold true. In other words, if $\{c_i\}$ are cycles generating the kth homology group, then for any corresponding real numbers, $\{a_i\}$, there exist a closed form, ω, such that

$$\oint_{c_i} \omega = a_i,$$

and this form is unique up to exact forms.

Stokes' theorem on smooth manifolds can be derived from Stokes' theorem for chains in smooth manifolds, and vice versa.Formally stated, the latter reads:

Theorem: *(Stokes' theorem for chains) If c is a smooth k-chain in a smooth manifold M, and ω is a smooth $(k-1)$- form on M, then*

$$\int_{\partial c} \omega = \int_c d\omega.$$

Underlying Principle

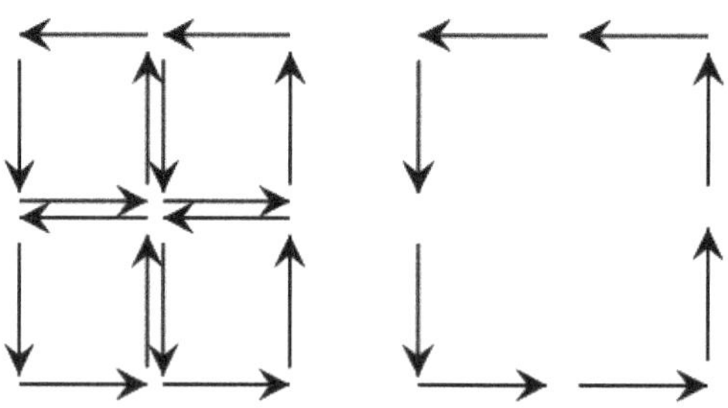

To simplify these topological arguments, it is worthwhile to examine the underlying principle by considering an example for $d = 2$ dimensions. The essential idea can be understood by the diagram on the left, which shows that, in an oriented tiling of a manifold, the interior paths are traversed in opposite directions; their contributions to the path integral thus cancel each other pairwise. As a consequence, only the contribution from the boundary remains. It thus suffices to prove Stokes' theorem for sufficiently fine tilings (or, equivalently, simplices), which usually is not difficult.

Generalization to Rough Sets

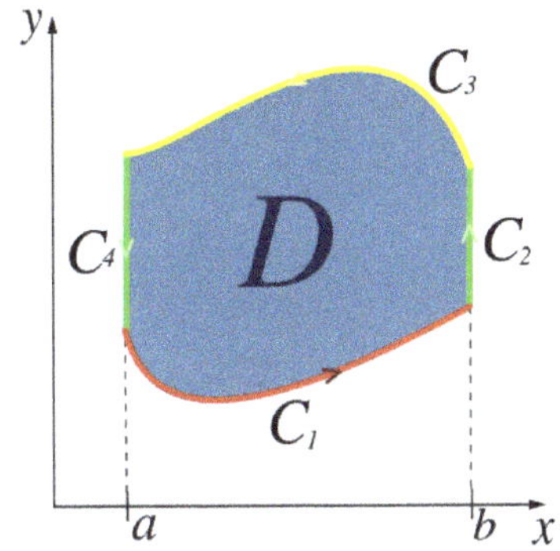

A region (here called *D*instead of Ω) with piecewise smooth boundary. This is a manifold with corners, so its boundary is not a smooth manifold.

The formulation above, in which Ω is a smooth manifold with boundary, does not suffice in many applications. For example, if the domain of integration is defined as the plane region between two x-coordinates and the graphs of two functions, it will often happen that the domain has corners. In such a case, the corner points mean that Ω is not a smooth manifold with boundary, and so the statement of Stokes' theorem given above does not apply. Nevertheless, it is possible to check that the conclusion of Stokes' theorem is still true. This is because Ω and its boundary are well-behaved away from a small set of points (a measure zero set).

A version of Stokes' theorem that extends to rough domains was proved by Whitney.Assume that D is a connected bounded open subset of R^n. Call D a *standard domain* if it satisfies the following property: There exists a subset P of ∂D, open in ∂D, whose complement in ∂D has Hausdorff $(n - 1)$-measure zero; and such that every point of P has a *generalized normal vector*. This is a vector $\mathbf{v}(x)$ such that, if a coordinate system is chosen so that $\mathbf{v}(x)$ is the first basis vector, then, in an open neighborhood around x, there exists a smooth function $f(x_2, ..., x_n)$ such that P is the graph $\{ x_1 = f(x_2, ..., x_n) \}$ and D is the region $\{ x_1 < f(x_2, ..., x_n) \}$. Whitney remarks that the boundary of a standard domain is the union of a set of zero Hausdorff $(n - 1)$-measure and a finite or countable union of smooth $(n - 1)$-manifolds, each of which has the domain on only one side. He then proves that if D is a standard domain in R^n, ω is an $(n - 1)$-form which is defined, continuous, and bounded on $D \cup P$, smooth on D, integrable on P, and such that $d\omega$ is integrable on D, then Stokes' theorem holds, that is,

$$\int_P \omega = \int_D d\omega.$$

The study of measure-theoretic properties of rough sets leads to geometric measure theory. Even more general versions of Stokes' theorem have been proved by Federer and by Harrison.

Special Cases

The general form of the Stokes' theorem using differential forms is more powerful and easier to use than the special cases. The traditional versions can be formulated using Cartesian coordinates without the machinery of differential geometry, and thus are more accessible. Further, they are older and their names are more familiar as a result. The traditional forms are often considered more convenient by practicing scientists and engineers but the non-naturalness of the traditional formulation becomes apparent when using other coordinate systems, even familiar ones like spherical or cylindrical coordinates. There is potential for confusion in the way names are applied, and the use of dual formulations.

Kelvin–Stokes' Theorem

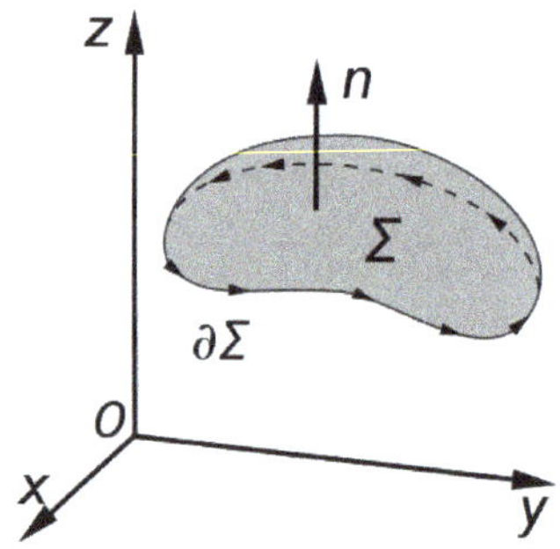

An illustration of the Kelvin–Stokes' theorem, with surface Σ, its boundary ∂Σ and the "normal" vector *n*.

This is a (dualized) (1 + 1)-dimensional case, for a 1-form (dualized because it is a statement about vector fields). This special case is often just referred to as *Stokes' theorem* in many introductory university vector calculus courses and as used in physics and engineering. It is also sometimes known as the curl theorem.

The classical Kelvin–Stokes' theorem relates the surface integral of the curl of a vector field over a surface Σ in Euclidean three-space to the line integral of the vector field over its boundary, is a special case of the general Stokes' theorem (with $n = 2$) once we identify a vector field with a 1-form using the metric on Euclidean 3-space. The curve of the line integral, ∂Σ, must have positive orientation, meaning that ∂Σ points counterclockwise when the surface normal, n, points toward the viewer.

One consequence of the Kelvin–Stokes' theorem is that the field lines of a vector field with zero curl cannot be closed contours. The formula can be rewritten as:

Suppose $\mathbf{F}(P(x,y,z),Q(x,y,z),R(x,y,z))$ is defined in region with smooth surface Σ and has first-order continuous partial derivatives. Then

$$\iint_{\Sigma}\left(\left(\frac{\partial R}{\partial y}-\frac{\partial Q}{\partial z}\right)dy\,dz+\left(\frac{\partial P}{\partial z}-\frac{\partial R}{\partial x}\right)dz\,dx+\left(\frac{\partial Q}{\partial x}-\frac{\partial P}{\partial y}\right)dx\,dy\right)$$
$$=\oint_{\partial\Sigma}\left(P\,dx+Q\,dy+R\,dz\right),$$

where *P*, *Q* and *R* are the components of F, and $\partial\Sigma$ is boundary of region with smooth surface Σ.

Green's Theorem

Green's theorem gives a relationship between the line integral of a two-dimensional vector field over a closed path in the plane and the double integral over the region it encloses. The fact that the integral of a (two-dimensional) conservative field over a closed path is zero is a special case of Green's theorem.

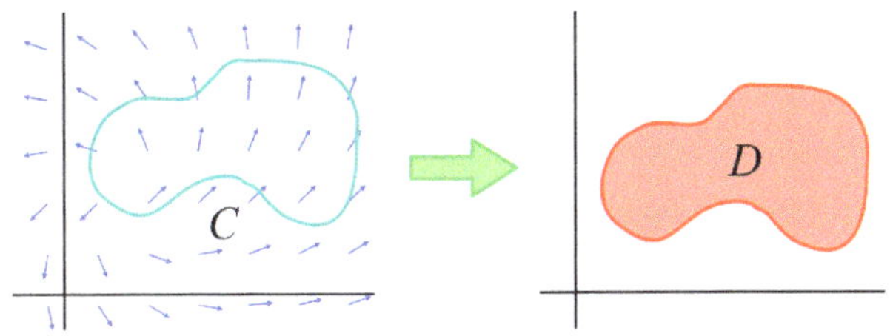

Green's theorem is itself a special case of the much more general Stokes' theorem. The statement in Green's theorem that two different types of integrals are equal can be used to compute either type: sometimes Green's theorem is used to transform a line integral into a double integral, and sometimes it is used to transform a double integral into a line integral.

Proof when D is a Simple Region

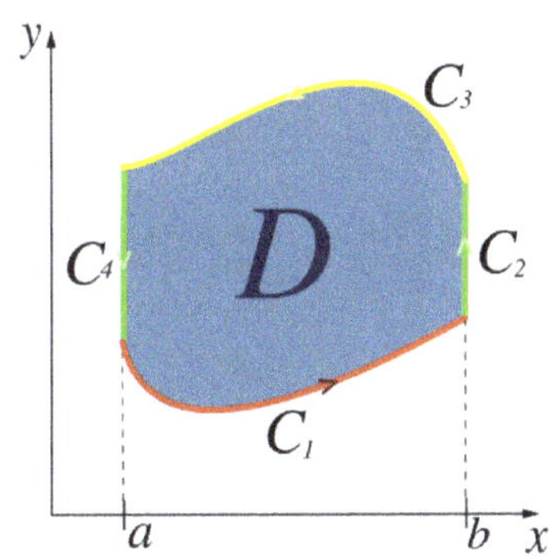

If D is a simple region with its boundary consisting of the curves C_1, C_2, C_3, C_4, half of Green's theorem can be demonstrated.

The following is a proof of half of the theorem for the simplified area D, a type I region where C_1 and C_3 are curves connected by vertical lines (possibly of zero length). A similar proof exists for the other half of the theorem when D is a type II region where C_2 and C_4 are curves connected by horizontal lines (again, possibly of zero length). Putting these two parts together, the theorem is thus proven for regions of type III (defined as regions which are both type I and type II). The general case can then be deduced from this special case by decomposing D into a set of type III regions.

If it can be shown that

$$\oint_C L\,dx = \iint_D \left(-\frac{\partial L}{\partial y}\right) dA$$

and

$$\oint_C M\,dy = \iint_D \left(\frac{\partial M}{\partial x}\right) dA$$

are true, then Green's theorem follows immediately for the region D. We can prove easily for regions of type I, and regions of type II. Green's theorem then follows for regions of type III.

Assume region D is a type I region and can thus be characterized, as pictured on the right, by

$$D = \{(x,y) \mid a \le x \le b, g_1(x) \le y \le g_2(x)\}$$

where g_1 and g_2 are continuous functions on $[a, b]$. Compute the double integral inabove equation:

$$\iint_D \frac{\partial L}{\partial y} dA = \int_a^b \int_{g_1(x)}^{g_2(x)} \frac{\partial L}{\partial y}(x,y)\,dy\,dx$$

$$= \int_a^b \{L(x, g_2(x)) - L(x, g_1(x))\} dx.$$

Now compute the line integral in (1). C can be rewritten as the union of four curves: C_1, C_2, C_3, C_4.

With C_1, use the parametric equations: $x = x$, $y = g_1(x)$, $a \le x \le b$. Then

$$\int_{C_1} L(x,y)dx = \int_a^b L(x, g_1(x))dx.$$

With C_3, use the parametric equations: $x = x$, $y = g_2(x)$, $a \le x \le b$. Then

$$\int_{C_3} L(x,y)dx = -\int_{-C_3} L(x,y)dx = -\int_a^b L(x, g_2(x))dx.$$

The integral over C_3 is negated because it goes in the negative direction from b to a, as C is oriented positively (anticlockwise). On C_2 and C_4, x remains constant, meaning

$$\int_{C_4} L(x,y)dx = \int_{C_2} L(x,y)dx = 0.$$

Therefore,

$$\int_C L\,dx = \int_{C_1} L(x,y)dx + \int_{C_2} L(x,y)dx + \int_{C_3} L(x,y)dx + \int_{C_4} L(x,y)dx$$

$$= -\int_a^b L(x, g_2(x))dx + \int_a^b L(x, g_1(x))dx.$$

Combining above equations, we get regions of type I. A similar treatment yields for regions of type II. Putting the two together, we get the result for regions of type III.

Proof for Rectifiable Jordan Curves

We are going to prove the following

Theorem. Let Γ be a rectifiable, positively oriented Jordan curve in $\mathbf{R}^2$ and let $\mathbf{R}$ denote its inner region. Suppose that $A, B : \overline{R} \to \mathbf{R}$ are continuous functions with the property that A has second partial derivative at every point of R, B has first partial derivative at every point of R and that the functions $D_1B, D_2A : R \to \mathbf{R}$ are Riemann-integrable over R. Then

$$\int_{\Gamma} A dx + B dy = \int_{R} \left(D_1 B(x, y) - D_2 A(x, y)\right) d(x, y).$$

We need the following lemmas:

Lemma 1 (Decomposition Lemma). Assume Γ is a rectifiable, positively oriented Jordan curve in the plane and let R be its inner region. For every positive real δ, let $\mathcal{F}(\delta)$ denote the collection of squares in the plane bounded by the lines $x = m\delta, y = m\delta,$, where m runs through the set of integers. Then, for this δ, there exists a decomposition of $\overline{R}$ into a finite number of non-overlapping subregions in such a manner that

(i) Each one of the subregions contained in R , say $R_1, R_2, \ldots, R_k$, is a square from $\mathcal{F}(\delta)$.

(ii) Each one of the remaining subregions, say $R_{k+1}, \ldots, R_s$, has as boundary a rectifiable Jordan curve formed by a finite number of arcs of Γ and parts of the sides of some square from $\mathcal{F}(\delta)$.

(iii) Each one of the border regions $R_k \;, \ldots, R_s$ can be enclosed in a square of edge-length 2δ.

(iv) If Γ_i is the positively oriented boundary curve of R_i, then $\Gamma = \Gamma_1 + \Gamma_2 + \cdots + \Gamma_s$.

(v) The number $s - k$ of border regions is no greater than $4\left(\dfrac{\Lambda}{\delta} + 1\right)$, where Λ is the length of Γ.

Lemma 2. Let $\Gamma.$ be a rectifiable curve in the plane and let $\Delta_\Gamma(h)$ be the set of points in the plane whose distance from (the range of) $\Gamma.$ is at most h. The outer Jordan content of this set satisfies $\overline{c}\,\Delta_\Gamma(h) \le 2h\Lambda + \pi h^2$.

Lemma 3. Let $\Gamma.$ be a rectifiable curve in $\mathbf{R}^2$ and let f : range of $\Gamma \to \mathbf{R}$ be a continuous function. Then

$\left|\int_\Gamma f(x, y)dy\right|$ and

$\left|\int_\Gamma f(x, y)dx\right|$ are $\le \dfrac{1}{2}\Lambda\Omega_f$, where Ω_f is the oscillation of f on the range of Γ.

Now we are in position to prove the Theorem:

Proof of Theorem. Let ε be an arbitrary positive real number. By continuity of A, B and compactness of $\overline{R}$, given $\varepsilon > 0$, there exists $0 < \delta < 1$ such that whenever two points of $\overline{R}$ are less than $2\sqrt{2}\delta$ apart, their images under A, B are less than ε apart. For this δ, consider the decomposition given by the previous Lemma. We have

$$\int_\Gamma A dx + B dy = \sum_{i=1}^{k} \int_{\Gamma_i} A dx + B dy \;\; + \sum_{i=k+1}^{s} \int_{\Gamma_i} A dx + B dy.$$

Put $\varphi := D_1 B - D_2 A$.

For each $i \in \{1,\ldots,k\}$, the curve Γ_i is a positively oriented square, for which Green's formula holds. Hence

$$\sum_{i=1}^{k}\int_{\Gamma_i} Adx + Bdy = \sum_{i=1}^{k}\int_{R_i}\varphi = \int_{\bigcup_{i=1}^{k} R_i}\varphi.$$

Every point of a border region is at a distance no greater than $2\sqrt{2}\delta$ from Γ. Thus, if K is the union of all border regions, then $K \subset \Delta_\Gamma(2\sqrt{2}\delta)$; ; hence $c(K) \le \bar{c}\Delta_\Gamma(2\sqrt{2}\delta) \le 4\sqrt{2}\delta + 8\pi\delta^2$, by Lemma 2. Notice that

$\int_R \varphi - \int_{\bigcup_{i=1}^{k} R_i}\varphi = \int_K \varphi.$ This yields

$$\left|\sum_{i=1}^{k}\int_{\Gamma_i} Adx + Bdy \quad - \int_R \varphi\right| \le M\delta(1+\pi\sqrt{2}\delta) \text{ for some } M > 0.$$

We may as well choose δ so that the RHS of the last inequality is $< \varepsilon$.

The remark in the beginning of this proof implies that the oscillations of A and B on every border region is at most ε. We have

$$\left|\sum_{i=k+1}^{s}\int_{\Gamma_i} Adx + Bdy\right| \le \frac{1}{2}\varepsilon\sum_{i=k+1}^{s}\Lambda_i.$$

By Lemma 1(iii),

$$\sum_{i=k+1}^{s}\Lambda_i \le \Lambda + (4\delta)4\left(\frac{\Lambda}{\delta}+1\right) \le 17\Lambda + 16.$$

Combining these, we finally get

$$\left|\int_\Gamma Adx + Bdy \quad - \int_R \varphi\right| < C\varepsilon,$$

for some $C > 0$. Since this is true for every $\varepsilon > 0$, we are done.

Validity under Different Hypothesis

The hypothesis of the last theorem are not the only ones under which Green's formula is true. Another common set of conditions is the following:

The functions $A, B : \bar{R} \to \mathbf{R}$ are still assumed to be continuous. However, we now require them to be Fréchet-differentiable at every point of R. This implies the existence of all directional derivatives, in particular $D_{e_i}A =: D_iA, D_{e_i}B =: D_iB, i = 1,2$, where, as usual, (e_1, e_2) is the canonical ordered basis of $\mathbf{R}^2$. In addition, we require the function $D_1B - D_2A$ to be Riemann-integrable over R.

It suffices to prove this for squares which are contained in R and have sides parallel to the axes.

The proof then follows the lines of the method employed to prove the Cauchy-Goursat Theorem for triangles.

As a corollary of this, we get the Cauchy Integral Theorem for rectifiable Jordan curves:

Theorem (Cauchy): If Γ is a rectifiable Jordan curve in $\mathbf{C}$ and if f : closure of inner region of $\Gamma \to \mathbf{C}$ is a continuous mapping holomorphic throughout the inner region of Γ, then

$$\int_\Gamma f = 0,$$

the integral being a complex contour integral.

Proof. We regard the complex plane as $\mathbf{R}^2$. Now, define $u, v : \bar{R} \to \mathbf{R}$ to be such that $f(x+iy) = u(x,y) + iv(x,y)$. These functions are clearly continuous. It is well-known that u and v are Fréchet-differentiable and that they satisfy the Cauchy-Riemann equations: $D_1 v + D_2 u = D_1 u - D_2 v = \text{zero function}$.

Now, analysing the sums used to define the complex contour integral in question, it is easy to realize that

$$\int_\Gamma f = \int_\Gamma u\,dx - v\,dy \quad + i\int_\Gamma v\,dx + u\,dy,$$

the integrals on the RHS being usual line integrals. These remarks allow us to apply Green's Theorem to each one of these line integrals, finishing the proof.

Measure-theoretic Assumptions

Green's formula also holds when, besides continuity assumptions,

(i) The functions $D_i A, D_i B, i = 1, 2,$, are defined at every point of R, with the exception of a countable subset.

(ii) The function $D_1 B - D_2 A$ is Lebesgue-integrable over R.

Multiply-connected Regions

Theorem. Let $\Gamma_0, \Gamma_1, \ldots, \Gamma_n$ be positively oriented rectifiable Jordan curves in $\mathbf{R}^2$ satisfying

$$\begin{aligned} &\Gamma_i \subset R_0, && \text{if } 1 \le i \le n \\ &\Gamma_i \subset \mathbf{R}^2 \setminus \bar{R}_j, && \text{if } 1 \le i, j \le n \text{ and } i \ne j, \end{aligned}$$

where R_i is the inner region of Γ_i. Let

$$D = R_0 \setminus (\bar{R}_1 \cup \bar{R}_2 \cup \cdots \cup \bar{R}_n).$$

Suppose $p : \bar{D} \to \mathbf{R}$ and $q : \bar{D} \to \mathbf{R}$ are continuous functions whose restriction to D is Fréchet-differentiable. If the function

$$(x,y)\mapsto\frac{\partial q}{\partial e_1}(x,y)-\frac{\partial p}{\partial e_2}(x,y)$$

is Riemann-integrable over D, then

$$\int_{\Gamma_0}p(x,y)dx+q(x,y)dy-\sum_{i=1}^{n}\int_{\Gamma_i}p(x,y)dx+q(x,y)dy$$
$$=\int_D\left\{\frac{\partial q}{\partial e_1}(x,y)-\frac{\partial p}{\partial e_2}(x,y)\right\}d(x,y).$$

Relationship to Stokes' Theorem

Green's theorem is a special case of the Kelvin–Stokes' theorem, when applied to a region in the *xy*-plane:

We can augment the two-dimensional field into a three-dimensional field with a *z* component that is always 0. Write F for the vector-valued function $\mathbf{F}=(L,M,0)$. Start with the left side of Green's theorem:

$$\oint_C(L\,dx+M\,dy)=\oint_C(L,M,0)\cdot(dx,dy,dz)=\oint_C\mathbf{F}\cdot d\mathbf{r}.$$

The Kelvin–Stokes' theorem:

$$\oint_C\mathbf{F}\cdot d\mathbf{r}=\iint_S\nabla\times\mathbf{F}\cdot\hat{\mathbf{n}}\,dS.$$

The surface S is just the region in the plane D, with the unit normals $\hat{\mathbf{n}}$ pointing up (in the positive *z* direction) to match the "positive orientation" definitions for both theorems.

The expression inside the integral becomes

$$\nabla\times\mathbf{F}\cdot\hat{\mathbf{n}}=\left[\left(\frac{\partial 0}{\partial y}-\frac{\partial M}{\partial z}\right)\mathbf{i}+\left(\frac{\partial L}{\partial z}-\frac{\partial 0}{\partial x}\right)\mathbf{j}+\left(\frac{\partial M}{\partial x}-\frac{\partial L}{\partial y}\right)\mathbf{k}\right]\cdot\mathbf{k}=\left(\frac{\partial M}{\partial x}-\frac{\partial L}{\partial y}\right).$$

Thus we get the right side of Green's theorem

$$\iint_S\nabla\times\mathbf{F}\cdot\hat{\mathbf{n}}\,dS=\iint_D\left(\frac{\partial M}{\partial x}-\frac{\partial L}{\partial y}\right)dA.$$

Green's theorem is also a straightforward result of the general Stokes' theorem using differential forms and exterior derivatives:

$$\oint_C L\,dx+M\,dy=\oint_{\partial D}\omega=\int_D d\omega$$
$$=\int_D\frac{\partial L}{\partial y}dy\wedge dx+\frac{\partial M}{\partial x}dx\wedge dy=\iint_D\left(\frac{\partial M}{\partial x}-\frac{\partial L}{\partial y}\right)dx\,dy.$$

Relationship to the Divergence Theorem

Considering only two-dimensional vector fields, Green's theorem is equivalent to the two-dimensional version of the divergence theorem:

$$\iiint_V (\nabla \cdot \mathbf{F}) dV = \oiint_{\partial v} (\mathbf{F} \cdot \mathbf{n}) dS.$$

where $\nabla \cdot \mathbf{F}$ is the divergence on the two-dimensional vector field $\mathbf{F}$, and $\hat{\mathbf{n}}$ is the outward-pointing unit normal vector on the boundary.

To see this, consider the unit normal $\hat{\mathbf{n}}$ in the right side of the equation. Since in Green›s theorem $d\mathbf{r} = (dx, dy)$ is a vector pointing tangential along the curve, and the curve C is the positively oriented (i.e. anticlockwise) curve along the boundary, an outward normal would be a vector which points 90° to the right of this; one choice would be $(dy, -dx)$. The length of this vector is $\sqrt{dx^2 + dy^2} = ds$. So $(dy, -dx) = \hat{\mathbf{n}} ds$.

Start with the left side of Green's theorem:

$$\oint_C (L dx + M dy) = \oint_C (M, -L) \cdot (dy, -dx) = \oint_C (M, -L) \cdot \hat{\mathbf{n}} ds.$$

Applying the two-dimensional divergence theorem with $\mathbf{F} = (M, -L),$, we get the right side of Green's theorem:

$$\oint_C (M, -L) \cdot \hat{\mathbf{n}} ds = \iint_D (\nabla \cdot (M, -L)) \, dA = \iint_D \left(\frac{\partial M}{\partial x} - \frac{\partial L}{\partial y} \right) dA.$$

Area Calculation

Green's theorem can be used to compute area by line integral. The area of D is given by

$$A = \iint_D dA.$$

Then if we choose L and M such that $\frac{\partial M}{\partial x} - \frac{\partial L}{\partial y} = 1$, the area is given by

$$A = \oint_C (L dx + M dy).$$

Possible formulas for the area of D include

$$A = \oint_C x dy = -\oint_C y dx = \tfrac{1}{2} \oint_C (-y dx + x dy).$$

Jacobian Matrix and Determinant

In vector calculus, the Jacobian matrix is the matrix of all first-order partial derivatives of a vector-valued function. Specifically, suppose is a function. Such a function is given by m real-valued component functions, . The partial derivatives of all these functions with respect to the variables can be organized in an m-by-n matrix, the Jacobian matrix of, as follows: This matrix, whose entries are functions of, is also denoted by and . The Jacobian matrix is important because if the function F is differentiable at a point, which is a slightly stronger condition than merely requiring that all partial derivatives exist there, then the derivative of F at p is the linear transformation represented by the matrix . This linear transformation is the best linear approximation of the function F near the point p. In the case the Jacobian matrix will be a square matrix, and its determinant, a function of, is the Jacobian determinant of F. It carries important information about the local behavior of F and can be thought of as a local expansion factor for volumes; it is used when performing variable substitutions in multi-variable integrals, since it occurs prominently in the substitution rule for multiple variables.

Jacobian Matrix

The Jacobian Matrix is just a matrix that takes the partial derivatives of each element of a transformation. In general, the Jacobian Matrix of a transformation F, looks like this:

$$J = \begin{bmatrix} \frac{\partial F_1}{\partial x_1} \cdots \frac{\partial F_1}{\partial x_n} \\ \vdots \quad \ddots \quad \vdots \\ \frac{\partial F_m}{\partial x_1} \cdots \frac{\partial F_m}{\partial x_n} \end{bmatrix} \quad F_1, F_2, F_3 \ldots$$

are each of the elements of the output vector and $x_1, x_2, x_3 \ldots$ are each of the elements of the input vector.

So for example, in a 2 dimensional case, let T be a transformation such that $T(u,v) =< x, y >$ then the Jacobian matrix of this function would look like this:

$$J(u,v) = \begin{bmatrix} \frac{\partial x}{\partial u} & \frac{\partial x}{\partial v} \\ \frac{\partial y}{\partial u} & \frac{\partial y}{\partial v} \end{bmatrix}$$

This Jacobian matrix noticably holds all of the partial derivatives of the transformation with respect to each of the variables. Therefore each row contains how a particular output element changes with respect to each of the input elements. This means that the Jacobian matrix contains vectors that help describe how a change in any of the input elements affects the output elements.

To help illustrate making Jacobian matrices, let's do some examples:

Example 1:

What would be the Jacobian Matrix of this Transformation?

$$T(u,v) =< u \times cosv, u \times sinv > .$$

Solution:

$$x = u \times cos\, v \rightarrow \frac{\partial x}{\partial u} = cos\, v, \frac{\partial x}{\partial v} = -u \times sin\, v$$

$$y = u \times sin\, v \rightarrow \frac{\partial y}{\partial u} = sin\, v, \frac{\partial y}{\partial v} = u \times cos\, v$$

$$J(u,v) = \begin{bmatrix} \frac{\partial x}{\partial u} & \frac{\partial x}{\partial v} \\ \frac{\partial y}{\partial u} & \frac{\partial y}{\partial v} \end{bmatrix} = \begin{bmatrix} cos\, v & -u \times sin\, v \\ sin\, v & u \times cos\, v \end{bmatrix}$$

Example 2:

What would be the Jacobian Matrix of this Transformation?

$T(u,v) =< u, v, u^{v} >, u > 0$.

Solution:

Notice, that this matrix will not be square because there is a difference in dimensions of the input and output, i.e. the transformation is not injective.

$$x = u \rightarrow \frac{\partial x}{\partial u} = 1, \frac{\partial x}{\partial v} = 0$$

$$y = v \rightarrow \frac{\partial x}{\partial u} = 0, \frac{\partial x}{\partial v} = 1$$

$$z = u^{v} \rightarrow \frac{\partial y}{\partial u} = u^{v-1} \times v, \frac{\partial y}{\partial v} = u^{v} \times ln(u)$$

$$J(u,v) = \begin{bmatrix} \frac{\partial x}{\partial u} & \frac{\partial x}{\partial v} \\ \frac{\partial y}{\partial u} & \frac{\partial y}{\partial v} \\ \frac{\partial z}{\partial u} & \frac{\partial z}{\partial v} \end{bmatrix} = \begin{bmatrix} 1 & 0 \\ 0 & 1 \\ u^{v} \times v & u^{v} \times ln(u) \end{bmatrix}$$

Example 3:

What would be the Jacobian Matrix of this Transformation?

$T(u,v) =< tan(uv) >$.

Solution:

Notice, that this matrix will not be square because there is a difference in dimensions of the input and output, i.e. the transformation is not injective.

$$x = tan(uv) \rightarrow \frac{\partial x}{\partial u} = sec^2(uv) \times v, \frac{\partial x}{\partial v} = sec^2(uv) \times u$$

$$(u,v) = \left[\frac{\partial x}{\partial u} \quad \frac{\partial x}{\partial v} \right] = [sec^2(uv) \times v \; sec^2(uv) \times u]$$

Jacobian Determinant

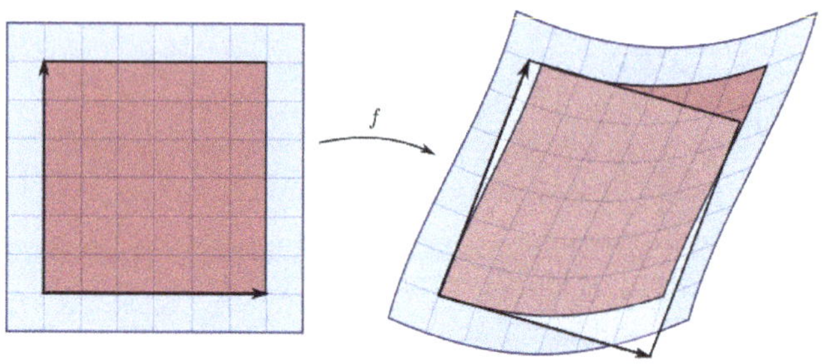

A nonlinear map $f : \mathbb{R}^2 \rightarrow \mathbb{R}^2$ sends a small square to a distorted parallelogram, shown in red. The Jacobian at a point gives the best linear approximation of the parallelogram near that point, shown as a translucent white.

If $m = n$, then **f** is a function from $\mathbb{R}^n$ to itself and the Jacobian matrix is a square matrix. We can then form its determinant, known as the Jacobian determinant. The Jacobian determinant is sometimes referred to as "the Jacobian".

The Jacobian determinant at a given point gives important information about the behavior of f near that point. For instance, the continuously differentiable function f is invertible near a point p ∈ $\mathbb{R}^n$ if the Jacobian determinant at p is non-zero. This is the inverse function theorem. Furthermore, if the Jacobian determinant at p is positive, then f preserves orientation near p; if it is negative, freverses orientation. The absolute value of the Jacobian determinant at p gives us the factor by which the function f expands or shrinks volumes near p; this is why it occurs in the general substitution rule.

The Jacobian determinant is used when making a change of variables when evaluating a multiple integral of a function over a region within its domain. To accommodate for the change of coordinates the magnitude of the Jacobian determinant arises as a multiplicative factor within the integral. This is because the n-dimensional dV element is in general a parallelepiped in the new coordinate system, and the n-volume of a parallelepiped is the determinant of its edge vectors.

The Jacobian can also be used to solve systems of differential equations at an equilibrium point or approximate solutions near an equilibrium point.

Generalizations of the Jacobian Determinant

The Jacobian determinant can be generalized also to the case where the dimension of the target differs from that of the domain. More precisely, let f, U , n, m, and y be as above:

- If $m < n$, the Jacobian of f at y is given by the square root of the determinant of $Df_y \cdot (Df_y)^t$ (where Df_y^t denotes the transpose of the matrix Df_y);

- If $m > n$, the Jacobian of f at y is given by the square root of the determinant of $(Df_y)^t \cdot Df_y$.

These generalizations play a key role respectively in the Coarea formula and Area formula.

An important characterization of the Jacobian is then given by the Cauchy Binet formula: $Jf(y)^2$ (is the sum of the squares of the determinants of all $n \times n$ minors of $Df|_y$.

$$\begin{bmatrix} cosv & -u & \times sin\, v \\ sinv & u & \times cos\, v \end{bmatrix} = u\cos^2 v + u\sin^2 v = u$$

Notice that, in an integral when changing from cartesian coordinates ($dxdy$) to polar coordinates ($drd\theta$), the equation is as such:

$$dxdy = r \times drd\theta = u \times dudv$$

It is easy to extrapolate, then, that the transformation from one set of coordinates to another set is merely

$$dC2 = det(J(T))dC1$$

where C1 is the first set of coordinates, det $(J(C1))$ is the determinant of the Jacobian matrix made from the Transformation T, T is the Transformation from C1 to C2 and C2 is the second set of coordinates.

It is important to notice several aspects: first, the determinant is assumed to exist and be non-zero, and therefore the Jacobian matrix must be square and invertible. This makes sense because when changing coordinates, it should be possible to change back.

Moreover, recall from linear algebra that, in a two dimensional case, the determinant of a matrix of two vectors describes the area of the parallelogram drawn by it, or more accurately, it describes the scale factor between the unit square and the area of the parallelogram. If we extend the analogy, the determinant of the Jacobian would describe some sort of scale factor change from one set of coordinates to the other. Here is a picture that should help:

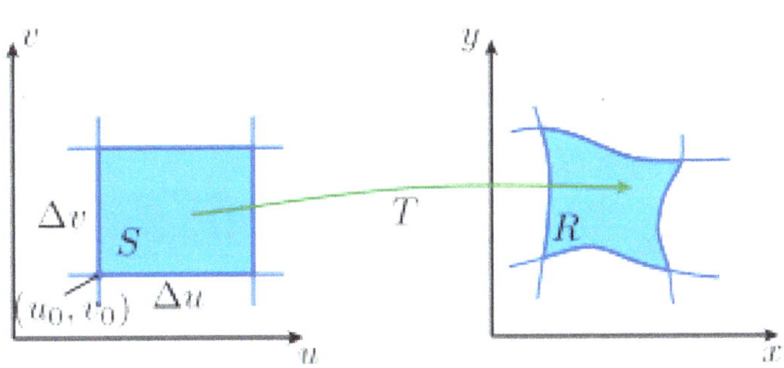

This is the general idea behind change of variables. It is easy to that the jacobian method matches up with one-dimensional change of variables:

$$T(u) = u^2 = x, \quad J(u) = \left[\frac{\partial x}{\partial u}\right] = [2u], \qquad du = |J(u)| \times du = 2u \times dx$$

Remember that when substituting, to be VERY careful. It should always be the case where

$$J(\vec{u}) \times d\vec{x} = d\vec{u}, \quad J(\vec{x}\ ) \times d\vec{u} = d\vec{x}\ .$$

Inverse

According to the inverse function theorem, the matrix inverse of the Jacobian matrix of an invertible function is the Jacobian matrix of the *inverse* function. That is, if the Jacobian of the function $\mathbf{f} : \mathbb{R}^n \rightarrow \mathbb{R}^n$ is continuous and nonsingular at the point $\mathbf{p}$ in $\mathbb{R}^n$, then $\mathbf{f}$ is invertible when restricted to some neighborhood of $\mathbf{p}$ and

$$\mathbf{J}_{\mathbf{f}^{-1}} \circ \mathbf{f} = \mathbf{J}_{\mathbf{f}}^{-1}.$$

Conversely, if the Jacobian determinant is not zero at a point, then the function is *locally invertible* near this point, that is, there is a neighbourhood of this point in which the function is invertible.

The (unproved) Jacobian conjecture is related to global invertibility in the case of a polynomial function, that is a function defined by n polynomials in n variables. It asserts that, if the Jacobian determinant is a non-zero constant (or, equivalently, that it does not have any complex zero), then the function is invertible and its inverse is a polynomial function.

Critical Points

If $\mathbf{f} : \mathbb{R}^n \rightarrow \mathbb{R}^m$ is a differentiable function, a *critical point* of $\mathbf{f}$ is a point where the rank of the Jacobian matrix is not maximal. This means that the rank at the critical point is lower than the rank at some neighbour point. In other words, let k be the maximal dimension of the open balls contained in the image of $\mathbf{f}$; then a point is critical if all minors of rank k of $\mathbf{f}$ are zero.

In the case where $1 = m = n = k$, a point is critical if the Jacobian determinant is zero.

Example 1:

Consider the function $\mathbf{f} : \mathbb{R}^2 \rightarrow \mathbb{R}^2$ given by

$$\mathbf{f}(x, y) = \begin{bmatrix} x^2 y \\ 5x + \sin y \end{bmatrix}.$$

Then we have

$$f_1(x, y) = x^2 y$$

and

$$f_2(x, y) = 5x + \sin y$$

and the Jacobian matrix of **F** is

$$\mathbf{J}_{\mathbf{f}}(x,y)=\begin{bmatrix}\dfrac{\partial f_1}{\partial x} & \dfrac{\partial f_1}{\partial y}\\ \dfrac{\partial f_2}{\partial x} & \dfrac{\partial f_2}{\partial y}\end{bmatrix}=\begin{bmatrix}2xy & x^2\\ 5 & \cos y\end{bmatrix}$$

and the Jacobian determinant is

$$\det(\mathbf{J}_{\mathbf{f}}(x,y))=2xy\cos y-5x^2.$$

Example 2: Polar-Cartesian Transformation

The transformation from polar coordinates (r, φ) to Cartesian coordinates (x, y), is given by the function $\mathbf{F}$: $\mathbb{R}^+ \times [0, 2\pi) \to \mathbb{R}^2$ with components:

$$x = r\cos\varphi;$$
$$y = r\sin\varphi.$$

$$\mathbf{J}_{\mathbf{F}}(r,\varphi)=\begin{bmatrix}\dfrac{\partial x}{\partial r} & \dfrac{\partial x}{\partial \varphi}\\ \dfrac{\partial y}{\partial r} & \dfrac{\partial y}{\partial \varphi}\end{bmatrix}=\begin{bmatrix}\cos\varphi & -r\sin\varphi\\ \sin\varphi & r\cos\varphi\end{bmatrix}$$

The Jacobian determinant is equal to r. This can be used to transform integrals between the two coordinate systems:

$$\iint_{\mathbf{F}(A)} f(x,y)dx\,dy=\iint_A f(r\cos\varphi, r\sin\varphi)r\,dr\,d\varphi.$$

Example 3: Spherical-Cartesian Transformation

The transformation from spherical coordinates (r, θ, φ) to Cartesian coordinates (x, y, z), is given by the function $\mathbf{F}\,\mathbb{R}^+ \times [0,\pi] \times [0,\ 2\pi) \to \mathbb{R}^3$ with components:

$$= r\sin\theta\cos\varphi;$$
$$y = r\sin\theta\sin\varphi;$$
$$z = r\cos\theta.$$

The Jacobian matrix for this coordinate change is

$$\mathrm{J}_{\mathrm{F}}(r,0,\varphi)=\begin{bmatrix}\dfrac{\partial x}{\partial r} & \dfrac{\partial x}{\partial \theta} & \dfrac{\partial x}{\partial \varphi}\\ \dfrac{\partial y}{\partial r} & \dfrac{\partial y}{\partial \theta} & \dfrac{\partial y}{\partial \varphi}\\ \dfrac{\partial z}{\partial r} & \dfrac{\partial z}{\partial \theta} & \dfrac{\partial z}{\partial \varphi}\end{bmatrix}=\begin{bmatrix}\sin\theta\cos\varphi & r\cos\theta\cos\varphi & -r\sin\theta\sin\varphi\\ \sin\theta\sin\varphi & r\cos\theta\sin\varphi & r\sin\theta\cos\varphi\\ \cos\theta & -r\sin\theta & 0\end{bmatrix}.$$

The determinant is $r^2 \sin \theta$. As an example, since $dV = dxdydz$ this determinant implies that the differential volume element $dV = r^2 sin\theta dr d\theta d\varphi$.. Unlike for a change of Cartesian coordinates, this determinant is not a constant, and varies with coordinates (r and θ).

Example 4:

The Jacobian matrix of the function $F : \mathbb{R}^3 \rightarrow \mathbb{R}^4$ with components

$$y_1 = x_1$$
$$y_2 = 5x_3$$
$$y_3 = 4x_2^2 - 2x_3$$
$$y_4 = x_3 \sin x_1$$

is

$$\mathbf{J}_{\mathbf{F}}(x_1, x_2, x_3) = \begin{bmatrix} \frac{\partial y_1}{\partial x_1} & \frac{\partial y_1}{\partial x_2} & \frac{\partial y_1}{\partial x_3} \\ \frac{\partial y_2}{\partial x_1} & \frac{\partial y_2}{\partial x_2} & \frac{\partial y_2}{\partial x_3} \\ \frac{\partial y_3}{\partial x_1} & \frac{\partial y_3}{\partial x_2} & \frac{\partial y_3}{\partial x_3} \\ \frac{\partial y_4}{\partial x_1} & \frac{\partial y_4}{\partial x_2} & \frac{\partial y_4}{\partial x_3} \end{bmatrix} = \begin{bmatrix} 1 & 0 & 0 \\ 0 & 0 & 5 \\ 0 & 8x_2 & -2 \\ x_3 \cos x_1 & 0 & \sin x_1 \end{bmatrix}.$$

This example shows that the Jacobian need not be a square matrix.

Example 5:

The Jacobian determinant of the function $\mathbf{F} : \mathbb{R}^3 \rightarrow \mathbb{R}^3$ with components

$$y_1 = 5x_2$$
$$y_2 = 4x_1^2 - 2\sin(x_2 x_3)$$
$$y_3 = x_2 x_3$$

is

$$\begin{vmatrix} 0 & 5 & 0 \\ 8x_1 & -2x_3 \cos(x_2 x_3) & -2x_2 \cos(x_2 x_3) \\ 0 & x_3 & x_2 \end{vmatrix} = -8x_1 \begin{vmatrix} 5 & 0 \\ x_3 & x_2 \end{vmatrix} = -40x_1 x_2.$$

From this we see that F reverses orientation near those points where x_1 and x_2 have the same sign; the function is locally invertible everywhere except near points where $x_1 = 0$ or $x_2 = 0$. Intuitively, if one starts with a tiny object around the point (1, 2, 3) and apply F to that object, one will get a resulting object with approximately $40 \times 1 \times 2 = 80$ times the volume of the original one, with orientation reversed.

Example 6:

Compute the following expression:

$$\iint (x^2 + y^2)dA$$

where dA is the region bounded by $y^2 - x^2 = 9,\ y^2 - x^2 = 16,\ 2xy = 14,\ 2xy = 6.$

Solution:

First, let's graph the region.

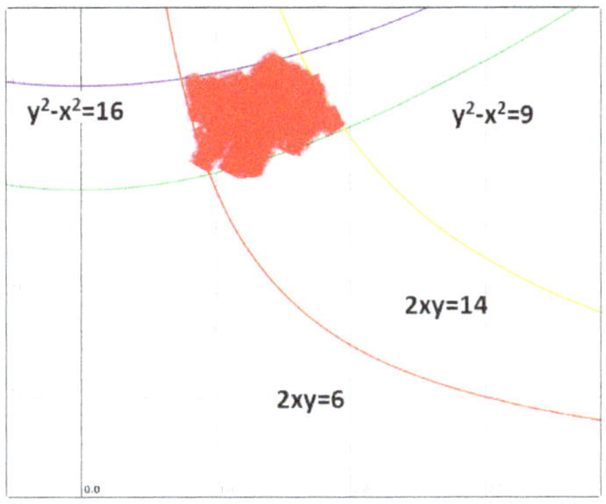

$$Let \quad v = 2xy, \quad u = y^2 - x^2$$

Why do we do this? Let's look at the bounds: they're from $y^2 - x^2 = 9$ to $y^2 - x^2 = 16$ and $2xy = 6$ and $2xy = 14$. It'd be quite a simple task if the integral looked something like this:

$$\int_9^{16}\int_6^{14} k(u,v)dudv$$

However, notice that in this case, instead of substituting x= and y=, we're substituing with u= and v=. Therefore the Jacobian will be written as $J(u,v)$ and the expression will be $dudv = J(u,v)dxdy$.

$$u = 2xy \rightarrow \frac{\partial u}{\partial x} = 2y, \frac{\partial u}{\partial y} = 2x$$

$$y^2 - x^2 \rightarrow \frac{\partial v}{\partial x} = -2x, \frac{\partial v}{\partial y} = 2y$$

$$J(u,v) = \begin{bmatrix} \frac{\partial u}{\partial x} & \frac{\partial u}{\partial y} \\ \frac{\partial v}{\partial x} & \frac{\partial v}{\partial y} \end{bmatrix} = \begin{bmatrix} 2y & 2x \\ -2x & 2y \end{bmatrix}$$

$dudv = |J(u,v)| \times dxdy \rightarrow \begin{bmatrix} 2y & 2x \\ 2x & -2y \end{bmatrix} \times dxdy = (4y^2 + 4x^2)dxdy$ The new region looks like The new region looks like this:

Therefore,

$$\iint (x^2+y^2)dA = \int_9^{16}\int_6^{14}(x^2+y^2)/(4y^2+4x^2)dudv = \int_9^{16}\int_6^{14} 1/4dudv = (16-9)\times(14-6)\times 1/4 = 14.$$

Example7:

Compute the following expression:

$$\iint \left(1/\sqrt{4x^2+9y^2}\right)dA$$

where dA is the region bounded by $(2x)^2+(3y)^2=36.$

Solution:

Again, let's graph the region.

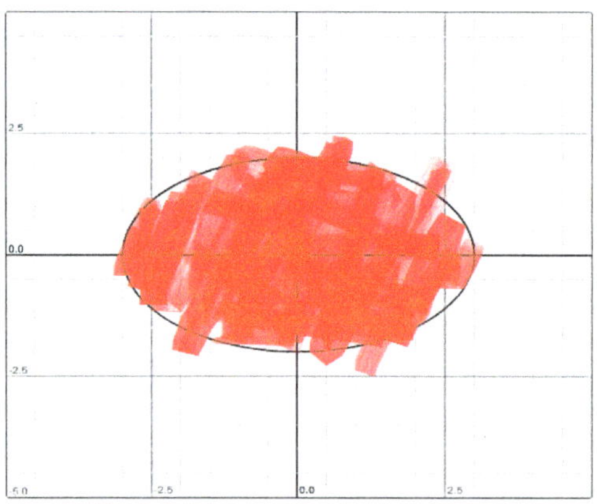

There are many ways to approach this, but to start, let's choose the change of variables to be $x = u/2,\ y = v/3$. Why?

$$(2x)^2+(3y)^2 = 36 \rightarrow u^2+v^2 = 36$$

The region now looks like this:

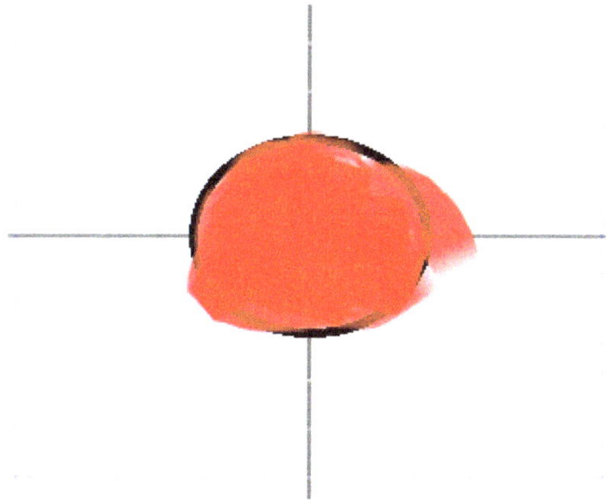

We've turned the region from an ellipse to a circle

However, notice that in this case, instead of substituting u= and v=, we're substituting with x= and y=. Therefore the Jacobian will be written as J(x,y) and the expression will be dxdy=J(x,y)dudv.

$$x = u/2 \rightarrow \frac{\partial x}{\partial u} = 1/2, \frac{\partial x}{\partial v} = 0$$

$$y = v/3 \rightarrow \frac{\partial y}{\partial u} = 0, \frac{\partial y}{\partial v} = 1/3$$

$$J(x,y) = \begin{bmatrix} \frac{\partial x}{\partial u} & \frac{\partial x}{\partial v} \\ \frac{\partial y}{\partial u} & \frac{\partial y}{\partial v} \end{bmatrix} = \begin{bmatrix} 1/2 & 0 \\ 0 & 1/3 \end{bmatrix}$$

$$dxdy = |J(x,y)| \times dudv = \begin{bmatrix} 1/2 & 0 \\ 0 & 1/3 \end{bmatrix} \times dudv = (1/6)dudv$$

$\iint \left(1/\sqrt{4x^2+9y^2}\right)dA = \iint 1/6 * 1\sqrt{u^2+v^2}\ dudv$ Now, let's change the coordinates to polar:

$$rsin\theta = u \quad ,rcos\theta = v,$$

$$\iint 1/6 \times 1/\sqrt{u^2+v^2}\ dA = 1/6 \times \iint 1/\sqrt{r^2} \times r \times drd\theta = 1/6 \times \int_0^{2\pi}\int_0^6 drd\theta = 1/6 \times 36 \times \pi = 6\pi$$

Example 8:

Compute the following expression:

$\iint \left(\sqrt{(x-y)(x+6y)}\right)dA$ where dA refers to the plane bounded by the points (0,0), (1,1), (7,0), (6,-1).

Solution:

So far, we've only discussed cases where separation of variables is used to make regions easier to integrate over. For this example, separation of variables will be used to make the integrand easier.

First, let's graph the region.

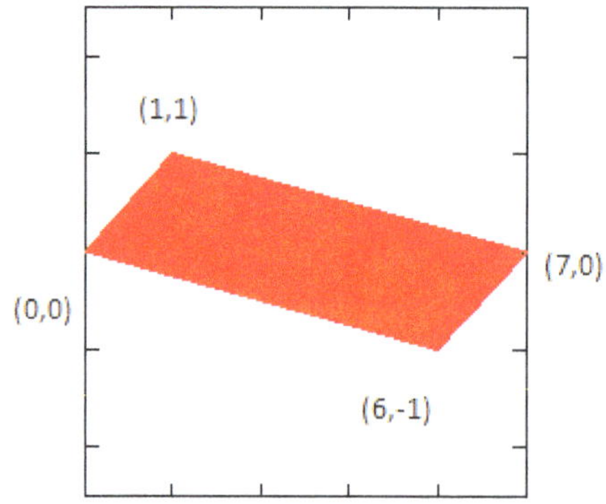

As previously stated, it would relatively simple to just do the double integral with this region. However, the integrand is the real problem. And so, to make things simpler, we'll set

$$u = x - y \rightarrow \frac{\partial u}{\partial x} = 1, \frac{\partial u}{\partial y} = -1$$

$$v = x + 6y \rightarrow \frac{\partial v}{\partial x} = 1, \frac{\partial v}{\partial y} = 6$$

Notice that in this case, we're back to substituting from the u= and v= side.

$$J(u,v) = \begin{bmatrix} \frac{\partial u}{\partial x} & \frac{\partial u}{\partial y} \\ \frac{\partial v}{\partial x} & \frac{\partial v}{\partial y} \end{bmatrix} = \begin{bmatrix} 1 & -1 \\ 1 & 6 \end{bmatrix}$$

$$dudv = |J(u,v)| \times dxdy = \begin{bmatrix} 1 & -1 \\ 1 & 6 \end{bmatrix} \times dxdy = (7)dxdy$$

Let's look at the transformed region. It just so happens that it made a 7 x 7 square!

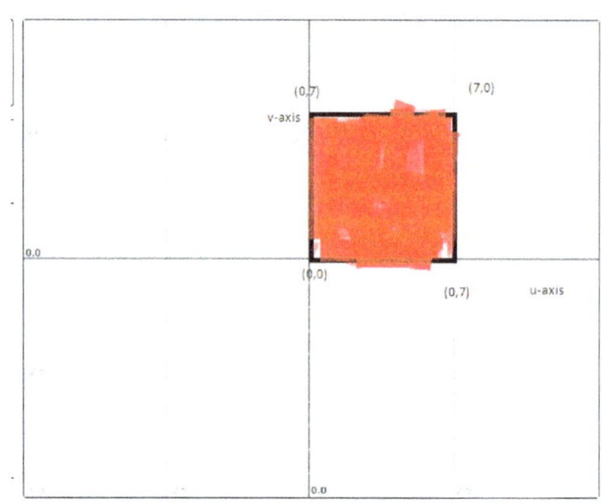

$$\iint\left(\sqrt{(x-y)(x+6y)}\right)dA=\int_0^7\int_0^7\sqrt{uv}dudv=7*3/2*3/2=63/4$$

Transforming Tangent Vectors

The major application of Jacobian Matrices comes from the fact that it is made of the partial derivatives of the transformation with respect to each original element. Therefore, when a function F is transformed by a transformation T, the tangent vector of F at a point is likewise transformed by the Jacobian Matrix.

Here is a picture to illustrate this.

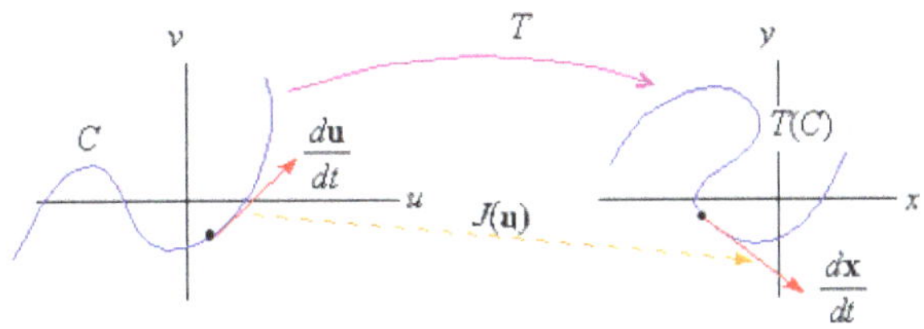

In other words, if we transform a function, we can find the new tangent vector at a transformed point.

Example:

$$<u,v>=<t^2,t^4>,\ T(u,v)=<u^3-v^3,3uv>=<x,y>$$

Find the velocity vector of $T(u,v)\ at\ t=1$.

Solution:

$$x=u^3-v^3\rightarrow\frac{\partial x}{\partial u}=3u^2,\frac{\partial x}{\partial v}=-3v^2$$

$$y=3uv\rightarrow\frac{\partial y}{\partial u}=3v,\frac{\partial y}{\partial v}=3u$$

$$J(x,y)=\begin{bmatrix}\frac{\partial x}{\partial u} & \frac{\partial x}{\partial v}\\ \frac{\partial y}{\partial u} & \frac{\partial y}{\partial v}\end{bmatrix}=\begin{bmatrix}3u^2 & -3v^2\\ 3v & 3u\end{bmatrix}$$

$$u'(t)=\begin{bmatrix}2t\\ 4t^3\end{bmatrix}$$

$$J(x,y)\times u'(t)\begin{bmatrix}3u^2 & -3v^2\\ 3v & 3u\end{bmatrix}\times\begin{bmatrix}2t\\ 4t^3\end{bmatrix}=\begin{bmatrix}3u^2\times 2t-3v^2\times 4t^3\\ 3v\times 2t+3u\times 4t^3\end{bmatrix}=\begin{bmatrix}3(t^2)^2\times 2t-3(t^4)^2\times 4t^3\\ 3\left(t^4\right)\times 2t+3\left(t^2\right)\times 4t^3\end{bmatrix}=\begin{bmatrix}6t^5-12t^{11}\\ 18t^5\end{bmatrix}$$

Therefore at t=1, the transformed tangent vector is:

$$\begin{bmatrix} -6 \\ 18 \end{bmatrix}$$

Linearization of Systems of Differential Equations

There is one last major application of Jacobians: approximating systems of Differential equations. Let's take the general expression:

$$\vec{x}' = f(\vec{x}, t)$$

Around a point, if $f(x,t)$ is differentiable, the system can be approximated to:

$$\vec{x}' \approx A\vec{x}, \quad A = J(\vec{x}, t_0)$$

The reason this works is similar to the idea behind one dimensional differentials:

$$dy \approx f'(x) \times dx$$

Why is this important? This can help us check stability for certain systems of differential equations. It is worth mentioning that this linearization works even when there are terms with t in them in the equation. This is because at very small intervals, those terms are small and don't affect the approximation.

Example 1:

Find the stability of the all the critical points of

$$\begin{bmatrix} x' \\ y' \\ z' \end{bmatrix} = \begin{bmatrix} z-x \\ 3/(2+x)-3y \\ 3y-z \end{bmatrix}$$

Solution:

To find the critical points, we set x'=y'=z'=0.

$$z = x, \quad y = 1/3x, \quad 3/(2+x) - x = 0 \rightarrow (x+3)(x-1) = 0, \quad (1,1/3,1) \quad (-3,-1,-3)$$

Now let's find the Jacobian at (1,1/3,1)

$$J(x', y', z') = \begin{bmatrix} -1 & 0 & 1 \\ -3/(2+x)^2 & -3 & 0 \\ 0 & 3 & -1 \end{bmatrix} = \begin{bmatrix} -1 & 0 & 1 \\ -1/3 & -3 & 0 \\ 0 & 3 & -1 \end{bmatrix}$$

The eigenvalues are -3.2055694304005904, -0.8972152847997048 ± 0.6654569511528134i.

Since the real parts are all negative, this critical point is locally stable.

Now let's find the Jacobian at (-3,-1,-3)

$$J(x',y',z') = \begin{bmatrix} -1 & 0 & 1 \\ -3/(2+x)^2 & -3 & 0 \\ 0 & 3 & -1 \end{bmatrix} = \begin{bmatrix} -1 & 0 & 1 \\ -3 & -3 & 0 \\ 0 & 3 & -1 \end{bmatrix}$$

The eigenvalues are -4, -2.5 ± 1.65831i. Because all of the real parts are negative, this critical point is locally stable.

Example 2:

Find the stability of the all the critical points of

$$\begin{bmatrix} x' \\ y' \\ z' \end{bmatrix} = \begin{bmatrix} z + 2y^2 + x \\ y - z^{3/2} \\ sin(z) \end{bmatrix}$$

Solution:

The only critical point in this case is (0,0,0)

Now, let's do the Jacobian.

$$J(x',y',z') = \begin{bmatrix} 1 & 2y & 1 \\ 0 & 1 & 3/2 * z^{1/2} \\ 0 & 0 & \cos(z) \end{bmatrix} = \begin{bmatrix} 1 & 0 & 1 \\ 0 & 1 & 0 \\ 0 & 0 & 1 \end{bmatrix}$$

The eigenvalue is just 1. Since this is positive, the critical point is unstable.

Example 3:

Find the stability of the all the critical points of

$$\begin{bmatrix} x' \\ y' \end{bmatrix} = \begin{bmatrix} y \\ -9sin(x) - y/5 \end{bmatrix}$$

Solution:

for critical points, we get that

$$y = 0, \quad x = \pi \times k, \quad k \subset N$$

Now, let's do the Jacobian.

$$J(x', y') = \begin{bmatrix} 0 & 1 \\ cos(x) & -1/5 \end{bmatrix}$$

$$Case1): x = 2\pi k \rightarrow J(x', y') = \begin{bmatrix} 0 & 1 \\ 1 & -1/5 \end{bmatrix}$$

$$Case2): x = \pi + 2\pi k \rightarrow J(x', y') = \begin{bmatrix} 0 & 1 \\ 1 & -1/5 \end{bmatrix}$$

For Case 1, the eigenvalues are -1.104987562112089 and 0.904987562112089 . Therefore, those points are saddles.

For Case 2, the eigenvalues are -0.1-0.99498743710662î and -0.1+0.99498743710662î. Therefore those critical points are spiral sinks.

Since this system is 2-d, it is possible to give a phase portrait and show you that this linearized system is accurate.

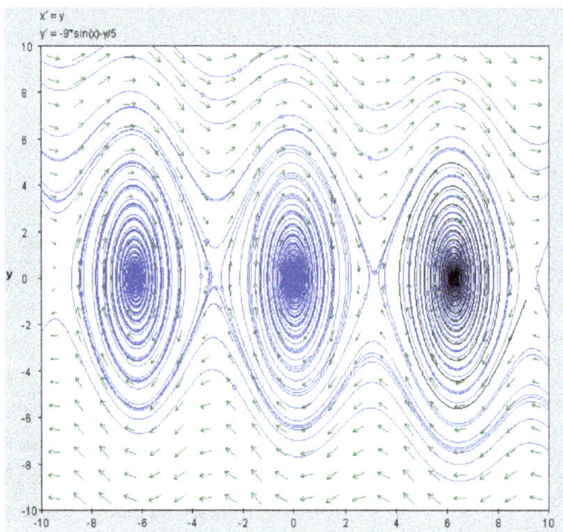

Example 4:

Find the stability of the all the critical points of

$$\begin{bmatrix} x' \\ y' \end{bmatrix} = \begin{bmatrix} 2x - y - x^2 \\ x - 2y + y^2 \end{bmatrix}$$

Solution:

For critical points, we have (0,0) and (1,1). Now, let's do the Jacobian.

$$J(x', y') = \begin{bmatrix} 2 - 2x & -1 \\ 1 & -2 + 2y \end{bmatrix}$$

$$Case1): x = 0, \quad y = 0 \rightarrow J(x', y') = \begin{bmatrix} 2 & -1 \\ 1 & -2 \end{bmatrix}$$

$$Case2): x = 1, \quad y = 1 \rightarrow J(x', y') = \begin{bmatrix} 0 & 1 \\ 1 & 0 \end{bmatrix}$$

In case 1, we get eigenvalues of 1.732 and -1.732. So, that point is a saddle.

In case 2, we get eigenvalues of -i and i. This is important because this is an exception to the linearization precision so far. Eigenvalues of -i and i would tell us that the critical point is a center. However, for this case only, the linearization will not be precise, and the actual graph may not have this critical point as a center.

Here's a phase diagram to show this:

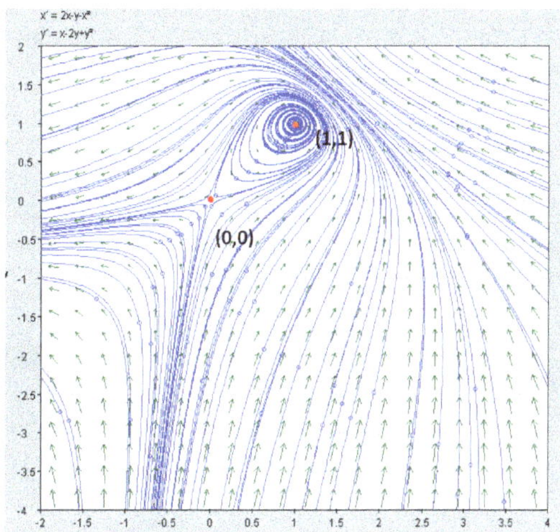

References

- Michael J. Crowe (1967). A History of Vector Analysis : The Evolution of the Idea of a Vectorial System. Dover Publications; Reprint edition. ISBN 0-486-67910-1
- Theresa, M. Korn; Korn, Granino Arthur. Mathematical Handbook for Scientists and Engineers: Definitions, Theorems, and Formulas for Reference and Review. New York: Dover Publications. pp. 157–160. ISBN 0-486-41147-8
- "Jacobian - Definition of Jacobian in English by Oxford Dictionaries". Oxford Dictionaries - English. Archived from the original on 1 December 2017. Retrieved 2 May 2018
- Katz, Victor J. (1979-01-01). "The History of Stokes› Theorem". Mathematics Magazine. 52 (3): 146–156. doi:10.2307/2690275. JSTOR 2690275
- Korn, Granino Arthur and Theresa M. Korn. Mathematical Handbook for Scientists and Engineers: Definitions, Theorems, and Formulas for Reference and Review. New York: Dover Publications. pp. 157–160. ISBN 0-486-41147-8
- "Jacobian - Definition of Jacobian in English by Oxford Dictionaries". Oxford Dictionaries - English. Archived from the original on 1 December 2017. Retrieved 2 May 2018

Permissions

All chapters in this book are published with permission under the Creative Commons Attribution Share Alike License or equivalent. Every chapter published in this book has been scrutinized by our experts. Their significance has been extensively debated. The topics covered herein carry significant information for a comprehensive understanding. They may even be implemented as practical applications or may be referred to as a beginning point for further studies.

We would like to thank the editorial team for lending their expertise to make the book truly unique. They have played a crucial role in the development of this book. Without their invaluable contributions this book wouldn't have been possible. They have made vital efforts to compile up to date information on the varied aspects of this subject to make this book a valuable addition to the collection of many professionals and students.

This book was conceptualized with the vision of imparting up-to-date and integrated information in this field. To ensure the same, a matchless editorial board was set up. Every individual on the board went through rigorous rounds of assessment to prove their worth. After which they invested a large part of their time researching and compiling the most relevant data for our readers.

The editorial board has been involved in producing this book since its inception. They have spent rigorous hours researching and exploring the diverse topics which have resulted in the successful publishing of this book. They have passed on their knowledge of decades through this book. To expedite this challenging task, the publisher supported the team at every step. A small team of assistant editors was also appointed to further simplify the editing procedure and attain best results for the readers.

Apart from the editorial board, the designing team has also invested a significant amount of their time in understanding the subject and creating the most relevant covers. They scrutinized every image to scout for the most suitable representation of the subject and create an appropriate cover for the book.

The publishing team has been an ardent support to the editorial, designing and production team. Their endless efforts to recruit the best for this project, has resulted in the accomplishment of this book. They are a veteran in the field of academics and their pool of knowledge is as vast as their experience in printing. Their expertise and guidance has proved useful at every step. Their uncompromising quality standards have made this book an exceptional effort. Their encouragement from time to time has been an inspiration for everyone.

The publisher and the editorial board hope that this book will prove to be a valuable piece of knowledge for students, practitioners and scholars across the globe.

Index

CPSIA information can be obtained
at www.ICGtesting.com
Printed in the USA
BVHW010459100619
550582BV00001B/49/P

9 781641 721400